W9-CVU-438

BIOLOGY

THE NEW YORK UNIVERSITY LIBRARY OF SCIENCE

BIOLOGY

EDITED BY

Samuel Rapport AND *Helen Wright*

ACADEMIC EDITORIAL ADVISER

HARRY A. CHARIPPER

Professor Emeritus of Biology,
New York University

NEW YORK • NEW YORK UNIVERSITY PRESS 1967
LONDON • UNIVERSITY OF LONDON PRESS LIMITED

Library of Congress Catalog Card Number: 67-10286.
Manufactured in the United States of America
Designed by Andor Braun

ACKNOWLEDGMENTS

"The Study of Life" from *The Forest and the Sea: A Look at the Economy of Nature and the Ecology of Man* by Marston Bates. Copyright © 1960 by Marston Bates. Reprinted by permission of Random House, Inc., and of Museum Press, Ltd., Publishers.

"Schleiden and Schwann: The Cell" from *The Coil of Life* by Ruth Moore. Copyright © 1961 by Ruth Moore. Reprinted by permission of Alfred A. Knopf, Inc.

"The Cell: Constituent Materials and Processes" abridged from *Life: An Introduction to Biology* by George Gaylord Simpson, Colin S. Pittendrigh, and Lewis H. Tiffany. Copyright © 1957 by Harcourt, Brace & World, Inc., and reprinted with their permission, and the permission of Routledge & Kegan Paul Ltd.

"Nutrition and Metabolism" from *Biology* by Karl von Frisch, translated by Jane M. Oppenheimer. Translation copyright © 1964 by Bayerischer Schulbuch-Verlag. Reprinted by permission of Harper & Row, Publishers.

"Odyssey of an Amoeba" from *Unresting Cells* by Ralph W. Gerard. Copyright © 1940, 1949 by Harper & Brothers. Reprinted by permission of Harper & Row, Publishers.

"Atoms into Life" from *The Atoms Within Us* by Ernest Borek. Copyright © 1961 by Columbia University Press. Reprinted by their permission.

"The Great Work" from *Charles Darwin* by Geoffrey West. Reprinted by permission of Yale University Press, and of David Higham Associates, Ltd.

"Gregor Mendel and His Work" by Hugo Iltis. Reprinted by permission of the Estate of the author. Copyright © 1942 by the American Association for the Advancement of Science.

"Evolution" from *Biology for the Modern World* by C. H. Waddington. Reprinted by permission of George G. Harrap & Co., Ltd.

"The Place of Genetics in Modern Biology" by George W. Beadle. The 11th Arthur Dehon Little Memorial Lecture at the Massachusetts Institute of Technology. Reprinted by permission of the Massachusetts Institute of Technology and of the author.

"Living Things" reprinted from *The Evolution of Life* by F. H. T. Rhodes by permission of Penguin Books Ltd.

"A Walk Through Time" from *Life of the Past* by George Gaylord Simpson. Copyright © 1953 by Yale University Press, and reprinted with their permission.

"The Story of a Leaf" reprinted with the permission of Charles Scribner's Sons from *The Living Past* by John C. Merriam. Copyright © 1930 Charles Scribner's Sons; renewal copyright © 1958 Lawrence C. Merriam, Charles W. Merriam, Malcolm L. Merriam.

"The Animal Community" from *Animal Ecology* by Charles Elton. Reprinted by permission of Sidgwick & Jackson Ltd. (Publishers).

"Strands of Dependence" from *Earth's Company* by Leslie Reid. Printed by permission of John Murray Ltd. (Publishers).

CONTENTS

V. The Ways of Living Things

FOREWORD

A PURPOSEFUL wandering in the forest of living things requires guidance based on mature judgment. Just being told that things biological depend on a combination of physical and chemical activities proceeding according to certain mathematical laws impresses one with some of the complexity of our being, and a desire to know more about it all.

The guided tour, as presented in this book about biology, should prove fascinating to the average person, scientist or layman. It begins with a concept of life in a complex world, stirred with zest by the competition for *lebensraum* and continuing to a revelation of the outstanding uniformity of the basic structure and functions of all things living—plant or animal. The puzzle of the finite difference between the animate and the inanimate becomes far from simple. Its solution is evasive and defies full resolution.

The admonition of "know thyself" infers a knowledge of biology—a knowledge of life and an appreciation of all things living, from the single cell structure to the complex multicelled organism—be it plant or animal. Man's quest for life in its essence has led from the obvious, the macroscopic, such as identifying the particular kind of bird, tree, or insect, to the microscopic, and now to the ultramicroscopic, for example, examining a special kind of cell or a particular part of the cell making a biochemical identification of the specific entity of inheritance in terms of DNA and/or RNA, the specific substance suspected of being the essence of life.

How close have we come? Will we control the factors of inheritance? Should we? These questions may be answered sometime in the future and will require the evaluation of society as a whole—in a cooperative unselfish effort to better our world, our life in it, and the integration of all living things for mutual advantage.

One cannot read and reread the aptly chosen chapters in this book on biology without appreciating the finely woven plan of life—the interrelation of the many living things—and without wondering "How? Why? And Where?" Above all, we ask that ever-puzzling question, What is life? How is life? What really makes you you and me me?

May you continue to ask yourself more questions, and have the interest to read in search of answers—better answers, satisfying answers. To a great extent the chapters in this volume may satisfy your interest, but if not, be brave and venture out beyond the confines of this book. Enjoy the adventure of searching for other writings, and above all, search your own mind, your own experience, your own observation; be prepared for the pleasant surprise of learning that you do know some things. This you will find helpful in assimilating your knowledge of the world you have been in for many years and the myriad of living things of which you have become aware.

Harry A. Charipper

INTRODUCTION

BIOLOGY, the study of living things, occupies a unique and central position between the natural and the social sciences. It leans heavily on the exact sciences. Without chemistry the study of the "internal environment," as Claude Bernard labeled it, would be nonexistent. Without microscopy the world of cellular structure and of bacteria and viruses would be invisible. Without knowledge of electrical impulses, the action of brain and nerves would remain incomprehensible. As the study of biology advances, its dependence on physical law increases; its ability to express its concepts in exact mathematical terms is a measure of its coming of age.

Biology, however, is itself the foundation on which another large sector of human knowledge and endeavor is built. The sociologist, the anthropologist, the aesthetician, and the artist are concerned with the actions and reactions of human beings; and homo sapiens, far from being unique and separate from other organisms, is made of the same basic materials and obeys the same biological laws. As more about these laws is discovered, more light is thrown on man's own nature, activities, and potentialities.

The beginnings of biology are hidden in prehistory. The hunter and the fisherman, the herdsman, and later the agriculturist, collected much practical knowledge of natural history. Some animals were easily domesticated, some grains unusually nutritious. The medicine man discerned that certain herbs or concoctions seemed helpful in the treatment of disease. From these activities sprang a body of knowledge, some of which has remained valuable to the present day (for example, prescription of digitalis to strengthen the pumping action of the heart). The needs of animal husbandry, agriculture, and medicine continue to exert a powerful stimulus to research in biological science.

Early biological study was practical in nature. The day of the "philosopher," the seeker after knowledge for its own sake, had not arrived. Five thousand years ago, the Egyptians and Babylonians made extended observations of plants and animals, but coherent systems were lacking. As in such other sciences as astronomy, the Greeks were the first to supply them. The late Professor L. L. Woodruff of Yale stated in a paper on the history of biology: "It is not an exaggeration to say that more than two thousand years ago the Greeks delineated the main subdivisions, the philosophical implications, and the social significance of biological science." Aristotle was the greatest of these Greek philosophers. He analyzed, classified, and insisted on direct observation. His researches dealt mainly with animals; his pupil Theophrastus wrote an *Enquiry on Plants* that entitles him to a place as one of the founders of botany. Hippocrates, the father of medicine, laid emphasis on the methods of induction and observation in the study of disease.

During the Roman period, physicians continued to contribute to the advancement of biological knowledge. Dioscorides, a Roman army surgeon, prepared a pharmacopoeia, a treatise on medicinally valuable herbs. Galen, a Roman physician of the second century, presented the anatomical and physiological knowledge of his day in a collection that became the standard work on medicine for over a millennium. These two works, together with the *Natural History* of Pliny the Elder, survived the dark ages to exert a profound and sometimes unfortunate influence on the biology of the early Renaissance. But gradually, with suitable deference to authority, the herbalists, the zoologists, and the encyclopedists made observations of their own, until a few men of genius broke the ancient bonds and embarked on the voyage which has resulted in modern biology. In 1543, the year of publication of Copernicus' *De Revolutionibus Orbium Coelestium,* Vesalius issued his great work on human anatomy *De Humani Corporis Fabrica.* An even greater scientific achievement was Harvey's tract on the circulation of the blood, which appeared in 1629. With these two works a scientific approach to the study of living things became possible. One further advance, the development of the microscope, should be noted. Without this invaluable instrument, it may truly be stated that modern biology would not exist.

Some of the great discoveries that followed are described in this book. From them has developed a body of knowledge of almost incredible scope and complexity. It includes such divisions as taxonomy, morphology, physiology, genetics, ecology, and palaeontology. There is much overlapping among these divisions. For example, the entomologist, who specializes in the study of insects, can attack his subject as anatomist, geneticist, or ecologist. Or he can limit himself to dipterology, the study of flies, or lepidoptery, the study of moths and butterflies. In any case, he hopes that his investigations will give rise to generalizations extending beyond his particular field. He realizes that great opportunities exist because biology, compared, for example, to physics, can be said to be in an early stage of development. Stimulated by recent discoveries in such areas as genetics and the study of the origin of life, biology may be on the brink of a period of growth comparable to that of physics in the first half of the twentieth century. But the discoveries of the future will be based on past achievements. No new advances in genetics can diminish the importance of Mendel, Morgan, Crick, and Watson. The biomicroscopist of the future must acknowledge his debt to Leeuwenhoek, Hooke, and Schwann. The experiments of Redi, Spallanzani, and Pasteur will strengthen the hands of investigators seeking the connection between life and nonlife. Like Einstein's modification of Newton's laws, any future modification of the law of organic evolution cannot dim the luster surrounding the names of Darwin and Wallace.

It has been customary to divide biology into two main sections: botany, the study of plants, and zoology, the study of animals. But biology is more than the sum of its parts. All living things, from the simplest to the most complex, share certain characteristics. They expend energy and must have food to replenish the supply; they reproduce; they respond to their enviroment. All are descended, through the processes of organic evolution, from similar rudimentary beginnings.

Each is dependent in greater or less degree on its fellow organisms. Living things are part of a system that is separate from inorganic matter. In recent years, the line of demarcation has become less distinct, but at least for purposes of study, it continues to exist. It is the characteristics that living organisms have in common that make up the science of biology.

The present book is addressed to laymen, and technical discussions have been held to a minimum. Scholarly apparatus, such as footnotes and references, have been omitted. In some cases, articles have been presented in somewhat abbreviated form. And the subject is of course so large that in a book of this size, many subdivisions can receive only the most cursory treatment or must be omitted completely.

BIOLOGY is one of the books in the series entitled THE NEW YORK UNIVERSITY LIBRARY OF SCIENCE. The series as a whole encompasses much of the universe of modern man, for that universe has been shaped in greatest measure by science, the branch of human activity whose name is derived from the Latin *scire*—to know.

I. Concepts of Biology

1. *Concepts of Biology*

The author of the following article has had an adventurous life in the biological sciences. A graduate of the University of Florida, he became a Ph. D. at Harvard in 1934. He has studied the malarial mosquito in Albania and the control of yellow fever in Colombia, both under the auspices of the Rockefeller Foundation. Drawing on these experiences, he has written a number of books which combine personal adventure with authentic contribution to biological knowledge. One of these books is The Forest and the Sea: A Look at the Economy of Nature and the Ecology of Man, *from which "The Study of Life" is taken. As a specialist in public health and human population problems, he joined the Foundation's New York office in 1949. In 1952, he became Professor of Biology at the University of Michigan. With other leading biologists, including the late Rachel Carson, Professor Bates is concerned with man's place in nature and with the possibly disastrous effects of man's activities on the environment. Professor Bates here elaborates on the nature and importance of biological science.*

THE STUDY OF LIFE

MARSTON BATES

PEOPLE OFTEN come to me with some strange animal they have found.

"What is it?" they ask.

Frequently I can't say—sometimes I get a despairing feeling of never knowing the answers to questions people ask. But at least I know where to look it up; or I know someone who is an expert on that kind of animal, so I can relay the question to him. And once in a while I know the answer.

"Oh," I say brightly, "that is a swallowtail butterfly, *Papilio cresphontes.*"

It is curious how happy people are to have a name for something, for an animal or plant, even though they know nothing about it beyond the name. I wonder whether there isn't some lingering element of word magic here, some feeling that knowing the name gives you power over the thing named—the sort of feeling that leads members of some savage tribes to conceal their personal names from all except their intimates. An enemy, learning their name, might be able to use this power for some evil purpose.

But other questions follow: "Where does it live?" and "What does it do?"

I explain that it is a tropical butterfly, common in Florida, which sometimes gets quite far north in the United States. The caterpillar lives on plants of the orange family, and north of Florida the butterfly is usually associated with prickly ash, which is a relative of the orange.

Almost inevitably there will come another question, "What good is it?"

I have never learned how to deal with this question. I am left appalled by the point of view that makes it possible. I don't know where to start explaining the world of nature that the biologist sees, in which "What good is it?" becomes meaningless. The question is left over from the Middle Ages; from a small, cozy universe in which everything had a purpose in relation to man. The question comes down from the days before Copernicus' theories removed the earth from the center of the solar system, before Newton provided a mechanism for the movements of the stars, before Hutton discovered the immensity of past time, before Darwin's ideas put man into perspective with the rest of the living world.

Faced with astronomical space and geological time, faced with the immense diversity of living forms, how can one ask of one particular kind of butterfly, "What good is it?"

Often my reaction is to ask in turn, "What good are you?"

Science has put man in his place; one among the millions of kinds of living things crawling around on the surface of a minor planet circling a trivial star. We can't really face the implications of this, and perhaps it is just as well—though I think humility is in general improving for the human character. A billion years into the past and a billion light years into space remain abstractions that we can handle glibly, but hardly realize. We remain important, you and I and all mankind. But so is the butterfly—not because it is good for food or good for making medicine or bad because it eats our orange trees. It is important in itself, as a part of the economy of nature.

The question ought to be not "What good is it?" but "What is its role in the economy of nature?" I like that phrase "the economy of nature," though there is a special word for the study of the interrelations of living things, *ecology*. Both words come from the Greek *oikos,* meaning "household"; both can have narrow and special meanings, but both can also be used broadly. Economics can be thought of as the ecology of man; ecology as the study of the economy of nature. This is one aspect of biology, one aspect of the study of life. It is thus also one aspect of science.

The word "science" covers a multitude of activities. We usually group these varied activities into three broad classes, which we call the physical sciences, the biological sciences, and the social sciences. This is reasonable. The physical sciences deal with matter and forces in the natural world. The points of view, the methods, the objectives, change somewhat when we add the element of life. These change again when we turn to man and have to deal with the added element of culture, of behavior governed by accumulated tradition and transmitted through speech and symbol systems.

Ultimately it may be possible to explain man in biological terms, and life in physico-chemical terms—to reduce all the complexities of poems and wars and bird songs to mathematical equations. But we are a long way from this and we find it most convenient now to operate rather differently at these different levels. Curiously, the physicists, sure that the ultimate answer will be theirs, tend to be a little scornful of the muddling biologists; and the biologists, convinced that man is an animal, are dubious about the fancy studies of the social scientists. But this shows only that scientists are human beings, and science another human activity.

The division into physical, biological, and social sciences looks logical enough, but it runs into all sorts of difficulties when we start to use it. What do we do with sciences like biochemistry and biophysics? And on the other hand, sciences like psychology and anthropology find themselves dealing with man as an animal as well as with man as a bearer of culture. Even with our present inadequate knowledge, it is clear that we are dealing with a continuum of natural events and that any division, however useful, is also artificial—reflecting the needs of the human mind rather than the realities of nature.

The difficulties—and the unrealities—increase when we turn to biology itself and to the problem of distinguishing among the different ways of studying life, which are the different kinds of biological sciences.

Sometimes I wonder whether there is any such thing as biology. The word was invented rather later—in 1809—and other words like "botany," "zoology," "physiology," "anatomy," have much longer histories and in general cover more coherent and unified subject matters. But while I doubt that biology has achieved a real existence yet, I am sure that botany, zoology, physiology, and anatomy ought not to exist. I would like to see the words removed from dictionaries and college catalogues. I think they do more harm than good because they separate things that should not be separated; because, however useful the words may have been in the past, they have now become handicaps to the further development of knowledge.

Words like "botany" and "zoology" imply that plants and animals are quite different things. They are different at one level, to be sure: anyone can tell a horse from an oak tree. But the differences rapidly become blurred when we start looking at the world through a microscope. There are many micro-organisms that botanists claim to be plants and zoologists to be animals, with equal plausibility. It was once logical to divide the objects in the world into three great classes

—animal, vegetable and mineral—but this distinction now is useful only in parlor games.

The similarities between plants and animals became more important than their differences with the discoveries that both were built up of cells with common basic characteristics, that plants, like animals, had sexual reproduction, and that their needs for nutrition and respiration were similar; and with the development of evolutionary theory, which showed that plants and animals were governed by the same kinds of evolutionary forces. Both are organisms. Unfortunately, "organism" remains a rather strange and pedantic word that has not really penetrated our basic vocabulary.

Plants and animals are different, of course. When we think of plants we first think of organisms with chlorophyl, organisms able to build up starches from carbon dioxide and water by using the energy of sunlight; they are the basic organisms in the economy of nature, on which all else depends. We also think of plants as fixed organisms; of animals as active, moving ones. And we think of plants in general as absorbing water and food; of animals as ingesting or "eating" it. There are exceptions to all of these. The fungi (mushrooms, molds, and the like) are clearly plants, but they lack chlorophyl and depend on other plants as much as animals do. As for movement, the slime molds, generally classed as plants, do a deal of creeping; and many kinds of animals in the sea, like the corals, are as immobile as any tree. I can't think of any plants that gobble their food, though the so-called carnivorous plants trap insects and digest them at leisure. Many animals, chiefly parasites, absorb food in a plantlike way.

These difficulties of definition are trivial. The most serious difficulties can be avoided by dividing all organisms not into two kingdoms, but into three—microbes, plants, and animals—because the microbes, in many ways, form a world just as distinctive as that of the visible plants and the visible animals.

But I still would object to dividing the study of living processes into botany, zoology, and microbiology, because, by any such arrangement, the interrelations within the biological community get lost. Corals cannot be studied without reference to the algae that live with them; flowering plants without the insects that pollinate them; grasslands without the grazing mammals. And at a different level, all protoplasm, all living stuff, shows much the same behavior: the problems of maintenance, growth, differentiation, reproduction, adaptation, evolution, are common to all life and can be studied most conveniently sometimes with one kind of organism, sometimes with another. The differences come out chiefly at the level of classification and cataloguing. This certainly is important enough, but it should not be allowed to swamp all other aspects of the study of life.

The case against physiology and anatomy is somewhat different. Anatomy is concerned with structure, physiology with function—which brings up the very old problem of the relation between structure and function, between how a thing is built and how it works. It is interesting that the first use of the word "biology" in the English language (it has an older history in German and French) was in a book published by an English surgeon in 1821, deploring not the separation of zoology and botany, but of anatomy and physiology. And this problem is still with us, only partially solved by labels like "functional anatomy."

Biology shares with medicine the tendency to give every possible kind of specialization a distinctive Greek-root label. This can be understood in medicine because the patient is presumably impressed by the thought that he is in the hands of an otolaryngologist or ophthalmologist or some other variety of learned specialist. It's part of the general medical love for big words and ritual, which I suspect has deep roots in our culture. The quacks and cultists have it as much as physicians in the tradition of scientific medicine. But this does not explain why people who study birds should want to be called ornithologists; insects, entomologists; grass, agrostologists; fungi, mycologists; and so on *ad infinitum* and *ad nauseam*.

I am not trying to argue against specialization. This, in the world of modern knowledge, is necessary. But generalization is also necessary if we are to fit our jigsaw pieces of information together into meaningful patterns. For this, I think the specialist should always be conscious of the relations of his particular subject and his particular point of view to the larger universe of knowledge, which is made more difficult if each way of specializing is called off as a distinct and independent science. There is an element of word magic here: entomology and limnology sound more like things-in-themselves than do insect biology and fresh water biology.

It is sometimes said that the unifying element in all of biology is the cell, that cells are the basic units of biology in the sense that atoms are the basic units of chemistry. I am dubious about this. Cellular organization, to be sure, is fundamental in visible organisms (plants and animals), but the world of the microbes is something else again.

Then we have the problem of viruses. Everyone knows about viruses: they cause polio, influenza, measles, and all sorts of other diseases. If the doctor doesn't know what is the matter with you, he is fairly safe in saying, "It's probably a virus." Yet everyone is puzzled about viruses. They certainly have no cellular structure in the ordinary sense. Some of them act like chemical solutions: they can be crystallized, liquefied again, and still show the properties of being alive. They are particulate, and it can be shown with the electron microscope that the different kinds of viruses come in different shapes and sizes. Some of them, at least, may be nothing more than giant molecules of a chemical stuff called nucleoprotein. One could plausibly argue that these nucleoproteins are the basic units of life.

But with this we have moved from biology to chemistry. Biology, chemistry, and physics form, or some day will form, a single, interdependent system of thought. People who call themselves chemists or physicists will probably eventually go furthest in revealing those tantalizing "secrets of life." But there are still plenty of things to be learned at the biological level. At this level, it seems to me that the most significant and general unit is not the cell, but the individual.

Living stuff is universally distributed in discrete packets, organized in the form of separate, individual organisms. These range from the single molecules of some virus particles, through all the varieties of noncellular organization of the microbes, to the coordinated aggregations of cells of an elephant, a whale, or a sequoia tree.

Yet there are difficulties with this concept of the individual. It is easy enough to identify individuals in a room full of people—to tell where one stops and the other starts. One can similarly identify individual marigolds in a flower bed, or individuals among the amoebae crawling across a microscope field. But it is not always so easy: the individual packets of life often maintain organic connections as colonies or clones, and I suspect it is sometimes a matter of rather arbitrary definition whether we call a given aggregate a "colony" or an "individual." We consider a coral clump to be a colony of individual polyps building a common skeleton, but a sponge is an individual organism. The tiny flagellate protozoans that are called *Volvox* form little spheres of dozens of cells that go tumbling along in a nicely coordinated fashion, but we call each sphere a colony. And what about a clump of bamboo or bananas, or a running grass?

There are other difficulties with the idea of the individual when we look at life over a span of time. All life is distributed in discrete packets; but all life is also continuous—the packets are momentary aspects of an ever-flowing stream through time. The amoeba or the bacterium splits in two. The protoplasm goes on, and we rather arbitrarily decide that the old individual is gone and that we have two new ones. With sex, we get something new: continuity depends on a fusion of part from two antecedent individuals, and the outward manifestation of the organization becomes discontinuous. The organization, the individual, must die and the continuity is through a germ-cell fragment that, for development, must unite with another fragment. Where is the individual here? I would say that the new individual starts at the moment when the two germ cells fuse—but this again could be called an arbitrary matter of definition.

Yet I think the individual is the nearest thing to an "objective" category that we have in biology. If I were planning a scheme for specialization within the biological sciences, I would take as my point of departure this idea of the individual. My first subdivision would depend on whether interest lay primarily in events inside the individual or outside. I would distinguish, in other words, between "skin-in" and "skin-out" biology.

This has a certain logic, because events inside the organism have to be studied by different methods from events outside the organism. Some people tend to be most interested in what goes on inside, in anatomy, digestion, circulation, and the like; others, in what the whole animal or plant does, how it behaves, how it is distributed around the world, how it lives.

To be sure, events inside the skin are by no means independent of those outside. An animal's behavior toward food, for instance, will depend on whether the animal is hungry or not, and hunger in turn depends on things going on in the digestive and nervous systems. All animal behavior depends to a large extent on sense perception—you can't react to a thing unless you can see it, smell it, taste it, feel it, or perceive it in some other way. The study of behavior thus involves the study of perception and sense organs, as well as nerve coordination, endocrine gland secretions, and all sorts of other things inside the skin. Inside and outside events in a plant are similarly related: you can't study growth, flowering, soil preference, cold tolerance, or anything of that sort without taking into consideration what is going on both inside and outside the plant. But this isn't saying that the skin, cortex, membrane, or whatever you want to call the boundary of the individual, is meaningless. It is saying that, for many things, you have to look on both sides of the boundary.

"Skin-in" biology, then, is primarily concerned with what makes the individual work, with the functioning of the different organ systems and with the ways in which they are built up of tissues and cells. "Skin-out" biology starts with this individual and is concerned with its relations to other individuals and to the varied aspects of the physical environment. If we start with the individual and go inside, we find ourselves concerned with such units as organs, tissues, cells, molecules. If we work in the other direction, we find that individuals of the same kind form populations, and that these populations can be grouped together into biological communities. When we talk about the economy of nature, we are talking about relationships among populations and individuals within these biological communities. I cannot make any very clear statement about man's place in nature, though my thinking is undoubtedly colored by my belief that man is a natural, rather than a supernatural, phenomenon. But whatever our beliefs, we are living with nature. And I think we can live more fully, more pleasantly, more productively, if we try to understand the world of nature. And in trying to understand nature, surely we also get new insights into ourselves.

II. The Cell

II. *The Cell*

Among his other achievements, Galileo, in 1610, made the first recorded observation with the compound microscope. The potentialities of this instrument for "a systematic study of the products of the body and of the general world of nature" were not immediately recognized. The work of the Englishman Hooke and the Dutchman Leeuwenhoek was to awaken scholars to a fantastic world of which they had previously been completely unaware. From these beginnings emerged the science of microbiology, on whose techniques research in cytology, bacteriology, and genetics largely depends. The development of one of these avenues resulted in the discovery of the fundamental biological unit, which Hooke named the cell. It is described in the following chapter from The Coil of Life, *by Ruth Moore. Miss Moore has become well known as the author of such authoritative works for the layman as* Man, Time and Fossils *and* The Earth We Live On. *Her selection deals with one of the great advances in biological science.*

SCHLEIDEN AND SCHWANN

RUTH MOORE

Each adds a little to our knowledge of Nature and from all the facts assembled arises grandeur.

ARISTOTLE

PERHAPS only a part of the unseen reality and strangeness that lay below the surface had been discovered by Bichat. If the organs were made up of tissues, was it possible that the tissues were composed of some more fundamental unit? Men of curiosity were constrained to wonder. And yet the discovery, when it finally came almost a third of a century later, was arrived at almost by indirection. The truth was largely unimaginable. It could not be put together out of the elements that men knew; it was beyond experience. Men might imagine gods and gnomes and unicorns, but who, other than a few visionaries, could dream that the body was made up of cells and organisms too small to be seen by the eyes, or felt by the hands, or apprehended in any other way by the senses?

The evidence of the cell had to be stumbled upon and pieced together over a period of nearly two hundred years. It was not until 1838 that two German scientists, Matthias Jacob Schleiden and Theodor Schwann, carried the work a little further and mobilized the facts so convincingly that men recognized that a new hidden unit had been discovered and a new direction taken.

Until the last years of the sixteenth century there was no way to look below the surface or beneath the structures laid bare by the scalpel. Only then did the Dutch spectacle-makers find that they could adjust two lenses in such a way as to enlarge an object under observation.

The great William Harvey (1578–1667) made use of his "perspicullum" to watch the beat of the hearts of wasps and flies. But almost another half-century passed before anyone began to use the microscope, as it was then named, to make a systematic study of the products of the body and of the general world of nature.

The microscope, with its ingenious arrangement of lenses and its ability to "overcome the infirmities of the senses," was exactly the kind of device to interest Robert Hooke (1635–1703).

Hooke had been born on the Isle of Wight on July 18, 1635. Orphaned at the age of eight, and sickly, he was apprenticed to the painter Sir Peter Lely. Only a small inheritance, to which he tenaciously clung, made it possible for him to enter Oxford when he was eighteen.

At Oxford in the 1650's was a group of young men interested in experimenting with the seemingly chaotic and unpredictable things that made up their environment, whether a pump to lift water or the movement of the planets. They were curious about many matters that had always been taken for granted. Among the group was the dashing Robert Boyle, fourteenth child of the "great" Earl of Cork. He engaged the poor and ugly but adroit young Hooke as his assistant. Hooke designed and built the air pump that is described in Boyle's first book, and he may have had more than a casual hand in the formulation of Boyle's law.

Pumps, springs, everything mechanical fascinated eighteen-year-old Hooke. He conceived the idea of using a spring to control the oscillation of a balance wheel in a watch, and he thus turned the erratic timepiece of the past into a precision instrument. He also seems to have developed the spiral spring, although he was soon involved in an acrimonious and continuing dispute—one of his many—to prove it.

Hooke was not the most agreeable of men, but there was no denying his inventiveness and orignality. In 1662, when the Royal Society was chartered, Hooke was named its curator. It was his duty to produce "three or four considerable experiments" for each of the weekly meetings.

For one of them the hard-worked curator brought in his own, improved compound microscope. To twentieth-century eyes it resembles a small fire extinguisher. A nozzle-like lens was attached to a cylinder elaborately ornamented with scrollwork. To light an object under examination, a candle was set up nearby and its light concentrated on the object by means of a reflector and a small convex-plano lens.

"By the help of the microscope," Hooke told the royal fellows, "there is nothing so small as to escape our inquiry."

He predicted to his willing though skeptical audience that it was probable, or at least not improbable, that by working with the microscope science might discover the composition of bodies, the various textures of "their matter," and perhaps even the manner of their "inward motion."

Hooke beamed his candle on the stinging hairs of a nettle, on the head of a fly, on the tufted gnat, on the edge of a knife, on the beard of a wild oat, and on hundreds of other specimens.

Among them was cork—common enough, certainly, and yet remarkably able to hold air in a bottle "without suffering the least bubble to pass" and quite "unapt to suck or drink in water."

One day Hooke sharpened his penknife until it was "as keen as a razor" and cut off a "good clear piece" of cork. This left the surface very smooth. He placed the smooth surface under the microscope and, moving it a little this way and that, thought that he could perceive it to be "a little porous." He could not be sure, although its lightness and springiness made porosity likely.

Hooke thought that with "further diligence" he could somehow make the cork structure more discernible. He took his sharpened knife and cut a very thin slice from the surface. The sliver seemed quite white, so he put it on a black object plate. "And casting the light on it with a deep plano-convex glass," said Hooke, "I could exceedingly plainly perceive it all to be perforated and porous, much like a Honey-comb."

The pores that sprang into view were regular and yet not completely regular; they also varied about as those in the honeycomb did. The similarity went even further. The cork had "very little solid substance in comparison to the empty cavity that was contain'd between the *Interstitia* (or walls, as I may so call them). The walls or partitions were near as thin in proportion to their pores as those thin films of wax in a honeycomb are to theirs."

"Next," said Hooke, naming for all time the fundamental unit of living matter, the cork was like honeycomb "in that those pores, or cells, were not deep but consisted of a great many little Boxes. . . .

"I no sooner discern'd these (which were indeed the first microscopic pores I ever saw, and perhaps that were ever seen, for I had not met with any Writer or Person, that had made any mention of them before this) but me thought I had with the discovery of them, presently hinted to me the true and intelligible reason of all the *Phaenomena of Cork.*"

Hooke bent over his honeycomb of cells. With a skillful hand he drew them from two different points of view, the one showing a fabric of little square compartments, the newly named cells, and the other looking down on them so that the cavities seemed rounder and their orderly arrangement in ranks or series was not visible. He put the two drawings side by side, labeling one "A" and the other "B," and set them off dramatically against a dark circular background. His apprenticeship to Lely had not been wasted.

If Hooke did not foresee the full implication of his epochal discovery, he understood well that it was important and promising. What he was seeing suggested the answer to the problem that had drawn him into the investigation of cork—why it floated and held air in a bottle so well. It seemed apparent that the cells were filled with air that was "perfectly enclosed" in each of the distinct little boxes.

"It seems very plain why neither the Water, nor any other Air can easily insinuate itself into them, since there is already within them an *intus existens,* and consequently, why the pieces of Cork become so good floats for Nets and stopples for Viols, or other close vessels," he explained.

The microscope showed that the whole mass was nothing but "an infinite company of small Boxes or Bladders of Air."

The air could be compressed, and that was why cork made such an excellent stopple and held so tight. Hooke was never known for his physical strength, for he constantly suffered from what today's physicians interpret as chronically inflamed sinuses, but he took several pieces of cork and squeezed them as hard as he could. He was able to "condense it into less than a twentieth part of its usual dimensions neer the Earth" simply by the use of his hands and without the aid of any "forcing Engine, such as Racks, Leavers, Wheels, Pullies, or the like."

Hooke divided off several lines of cork pores and counted them. There were usually about sixty of the small cells end to end in an eighteenth of an inch.

Not a brilliant mathematician, Hooke was unable to work out the curvilinear motions of the planets, and thus in his planetary studies he lost eternal fame to his rival Isaac Newton. But he had no difficulty with common arithmetic. There were more than 1,000 cells to an inch, and more than a million—1,166,400 to be exact—in a square inch, and 1,259,721,000 in a cubic inch.

Hooke was startled. It was incredible, a thing that could not have been believed if the microscope had not established it by ocular demonstration.

And these stupefyingly numerous little cells were not peculiar to cork. Other vegetables, Hooke found, had the same compartmented structure, and in some of them the cells were even smaller than in cork. Hooke examined the pith of the elder and the pulp or pith of the "cany" stalks of such vegetables as "carrets, daucus, Bur-docks, teasel, fearn and several kinds of reed." Even the pith that filled the stalk of a feather had the same cellular texture or "schematisme."

The cells that Hooke was examining so eagerly were completely walled. And yet he saw that the juices must pass in and out through their seemingly solid partitions, for the cells of living things were filled with juices, while intact cells in charred material were "empty of everything but Air."

How could this be? How could matter pass through the unbroken walls? Hooke decided to seek for the passage. He blew on the cells, hoping to force out their contents. This produced no results. Could it be that the vegetable pores had valves, like those of the animal heart, that might give passage "to the contain'd fluid juices one way and shut themselves, and impede the passage of such liquids back again?" Try as he would, Hooke could find no valves. His failure to discover the entry and exit did not mean, he hastened to add, that Nature did not have some kind of a contrivance "to bring her designs and ends to pass." With a better microscope he was hopeful that it could be found.

Hooke reported his observations on cork and a gamut of other things, ranging up to the craters on the moon, in a book that he called the *Micrographia or Some Physiological Descriptions of Minute Bodies Made by Magnifying Glasses with Observations and Inquiries Thereupon.* It was dedicated to the King—"I do most humbly lay this small present at your Majesties Royal feet"—and published in 1665.

The book aroused intense interest, even at a time when any attention at all might have seemed impossible. In 1665 the Great Plague was at its dread peak and people were dying by the thousands. The next year London was swept by the Great Fire. And yet amid all the ravages of disease and flame, the invisible had become visible, and the revelation was a startling one. Samuel Pepys sat up until two o'clock in the morning reading the *Micrographia*. He called it "the most ingenious book that ever I read in my whole life."

Hooke was only twenty-nine, but his reputation was established. And so was the cell. Men knew at last that beneath everything they saw lay tiny cells or compartments.

Other men of science and curiosity were becoming interested in the newfangled instrument that let men see things that always before had been "buried from their eyes." Pepys, no scientist but not a man to miss anything new going on about him, paid five pounds for one of the instruments, which he noted was "a great price for a curious bauble."

In 1673, eight years after the *Micrographia* had been published, the Royal Society received a letter from one of its Dutch correspondents, Reinier de Graaf. "I am writing to tell you," he said, "that a certain most ingenious person here, named Leeuwenhoek has devised microscopes which far surpass those which we have hitherto seen. . . ."

De Graaf enclosed a letter from the "ingenious" Leeuwenhoek. It indeed more than merited the adjective, for it described some unusual microscopic observations on mold, on the sting, mouth parts, and eyes of the bee, and on the louse. The Royal Society was sufficiently impressed to invite Leeuwenhoek to send them any further observations he might make.

Anton van Leeuwenhoek (1632–1723) had not been brought up for such scientific activity and formal correspondence. His only training was as a draper. And when he began to make the fine microscopes of which De Graaf wrote, he was busy at the dual and unrelated occupations of selling ribbons and buttons and seeing to the proper maintenance of the town hall. He was chamberlain to the burgomasters, a post that made him responsible for the lay-ing of fires in the towered Delft town hall and keeping it neat and tidy.

The Delft draper and custodian was at the same time a man of rare curiosity, persistence, and confidence. When he began to make microscopes—in actuality they were magnifying glasses—he not only ground his own lenses to a new perfection, but even extracted the gold and silver in which he mounted some of the plates.

Leeuwenhoek felt overwhelmed at the notice of the Royal Society of London, certainly one of the world's most distinguished bodies of philosophers, as they were then called. He knew none of the Latin in which he should properly have addressed so learned a body. Nevertheless, Leeuwenhoek sent along some of his drawings of the bee and explained in his nonliterary Dutch: "I beg that those Gentlemen to whose notice these may come please to bear in mind that my observations and thoughts are the outcome of my own unaided impulse and curiosity alone; for beside myself in our town there be no philosophers who practice this art; so pray take not amiss my poor pen and the liberty I have taken in setting down my random notions."

Leeuwenhoek was forty-one and had already made dozens of microscopes. Instead of changing his specimens, he kept them fixed under his lens and made additional microscopes for viewing the vast and varied array that he wanted to study. Perhaps making a new instrument was as easy as gluing an object to the point of a needle and getting it before his glass.

The next year Leeuwenhoek was walking beside a lake near Delft when he noticed trailing clouds of green stuff in the water. The country folk said that the growth was produced by the dew each summer. Leeuwenhoek collected a little bit of the slimy green and mounted it under his microscope.

Green streaks that wound about almost like serpents leaped out at him; there also were many small round green globules, and among them were little "animalcules," some roundish and some oval.

"On the last two," said Leeuwenhoek, "I saw two little legs near the head and two little fins at the hindmost end of the body. . . . And the motion of most of these animalcules in the water was so swift and so various, upwards, downwards and round about, that 'twas wonderful to see: and I judge that some of the little creatures were about a thousand times smaller than the smallest ones [mites] that I have ever yet seen upon the rind of cheese, in wheaten flour, mould and the like."

Neither the jungles of Africa nor the imaginings of mythology could equal the fantastic forms and behavior of the little monsters that swarmed, tumbled, and writhed in this drop of pond water. Leeuwenhoek stared in disbelief at his first sight of the one-celled animals that were later to be classified as members of the phylum Protozoa. But he was not suffering from illusions: they did not vanish with a blink of the eye; they were as real as his microscope. Was it possible that other water could conceal the same kind of outlandish creatures?

In September 1675 Leeuwenhoek happened to take a look at some rain water that had been standing for a few days in a container lined with Delft blue glaze. The water too was full of little animals, astoundingly little ones. What could they be compared to? What measurement could convey even an idea of their minuteness? About the smallest animal of which Leeuwenhoek could think, an animal even smaller than the cheese and other mites, was the water flea that Jan Swammerdam had portrayed and studied. It was just large enough to be seen with the naked eye. His animals, Leeuwenhoek calculated, were about ten thousand times smaller than the barely visible water flea.

Leeuwenhoek studied his animalcules day after day. He could describe their behavior as clearly as he could that of his cat.

"When these animalcules bestirred themselves, they sometimes stuck out two little horns which were continually moved, after the fashion of a horse's ears. The part between these little horns was flat, their body being roundish, save only that it ran somewhat to a point at the hind end; at which pointed end it had a tail, near four times as long as the whole body, and looking as thick when viewed through the microscope as a spider's web. At the end of this tail there was a pellet, of the bigness of one of the globules of the body; and this tail I could now perceive to be used by them for their movements in very clear water.

"These little animals were the most wretched creatures that I have ever seen; for when, with the pellet they did but hit on any particles or little filaments (of which there are many in the water, especially if it hath but stood some days), they stuck intangled in them; and they pulled their body out into an oval and did struggle by strongly stretching themselves to get their tail loose; whereby their whole body then sprang back towards the pellet of the tail, and their tails then coiled up serpent-wise, after the fashion of a copper or iron wire that, having been wound close around a round stick, and then taken off, kept all its windings.

"This motion of stretching out and pulling together the tail continued; and I have seen several hundred animalcules caught fast by one another in a few filaments, lying within the compass of a coarse grain of sand." [1]

Other equally remarkable animals flashed through the drop of water. Some waved "incredibly thin little feet or legs"—Leeuwenhoek's description of the hairlike appendages that are now called cilia. Others, about twice as long as they were broad, sometimes stood still "twirling themselves round as 'twere, with a swiftness such as you see in a whip-top a-spinning." And there were innumerable others, some monstrous in size as compared to the tiniest ones.

Leeuwenhoek was not a man to speculate, but he did ask essential questions. Where did this freakish menagerie come from? Did the animalcules come down from the sky with the rain?

On the 26th of May, 1676, a heavy rain fell in Delft. During a brief let-up Leeuwenhoek took a clean glass and caught some of the water as it came off his slate roof. He hurried in with it and put a drop under his microscope. A few animals swam about in it, but Leeuwenhoek, in think-

1. Undoubtedly vorticella, one-celled ciliated animals with bell-shaped bodies. Leeuwenhoek did not see that the tails might be stalks, and thus he thought the animals were trying to break out of an entanglement.

ing hard about it, realized that they might have been bred in his leaden gutters.

The rain was continuing, so the careful Dutchman carried a large porcelain dish into his courtyard and sat it on a tub high enough to prevent any mud from splashing into it. He collected some more water. This time he found no animals, though there were a lot of earthy particles in this water. Then the animals did not rain down from the skies.

But keep the water a few days, or let the wind ruffle it, and Leeuwenhoek found as many as a thousand animals in a drop, "a-wallowing on their back as well as their belly, and all a-rolling." The animals turned up in river water, in sea water, and even in the cold palatable water that Leeuwenhoek drew from the fifteen-foot well in his yard. And they swarmed in pepper water.

Leeuwenhoek chanced upon the last. In the course of trying everything under his microscope, he decided to have a look at pepper. He had always wondered what caused the "hotness of power whereby pepper affects the tongue." It was difficult to work with dry grains, so he put some to soak. About three weeks later when he got around to having a look at the mixture, the sight was flabbergasting. The pepper water fairly teemed with such a multitude of animals as he had not seen before. Among them were some extremely thin little tubes, infinitely smaller than the smallest of the animalcules. Leeuwenhoek was looking at bacteria, although he did not know or suspect that he was seeing for the first time the invisible group that can do both untold ill and good to man. Leeuwenhoek only marvelled at the size and number of the rods. They outnumbered many times the six to eight thousand animalcules in a drop of water.

A full report on this newly discovered world and all its wonders had to be dispatched to the Royal Society. Leeuwenhoek had his laboratory notes copied out in a fine hand on seventeen folio pages and sent them off to London.

The effect was sensational. As many animals in a few drops of water as there were on the earth! Another world lying concealed in the universe that men had always known! The society was staggered. Leeuwenhoek's work bore all the marks of exactitude, but what he was saying went beyond what anyone could accept without the most rigorous proof. Leeuwenhoek was requested to supply more data, and Hooke, who had been made secretary of the society in 1677, was asked to see if he himself, as a man who had worked with microscopes, could confirm the fantastic report from Holland.

Hooke's microscope had been neglected for several years, "I having been by other urgent occupations diverted," but he quickly got it out and made up some pepper water as Leeuwenhoek had done. And there were the animals in all their multitudes and their freakishness.

"It seems very wonderfull," Hooke told the society, "that there should be such an infinite number of animalls in soe imperceptible a quantity of matter; that these animalls should be soe perfectly shaped and indeed with such curious organs of motion as to be able to move nimbly, to turne, stay, accelerate & retard their progresse at pleasure. And it was not less surprising to find that these were gygantick monsters in comparison to a lesser sort which almost filled the water."

Thus Hooke confirmed Leeuwenhoek's discovery of both the protozoa and bacteria, although neither knew the strange new population in these still unthought-of terms.

People anxiously flocked to see the marvel, among them His Majesty Charles II, the founder and patron of the Royal Society. Hooke was also asked to demonstrate the animals before a formal session of the society. The experiments that he set up for the meetings of November 1 and 8, 1677, ran into difficulties, but at the next weekly meeting, on November 15, with the aid of a better microscope and thinner glass pipes to contain the drop of pepper water, all went well. Sir Christopher Wren and others named to a special committee saw and affirmed that the little animals were swimming to and fro, exactly as Leeuwenhoek had said. The unlettered Dutchman had made one of the greatest of discoveries. This was evident even then.

Leeuwenhoek did not rest on this dazzling affirmation. If the animalcules were present in all water except that which had just fallen from the heavens, perhaps they also lodged in the human body. Leeuwenhoek reasoned that if they were in the body, they would find their way to the mouth.

Each morning all through his life he had rubbed his teeth with salt and cleaned them with a toothpick. Nevertheless when he examined his teeth with a magnifying glass he found around them a little white matter "as thick as 'twere batter." He scraped off a bit and mixed it with rain water in which there were no animalcules, and with a little spittle.

As his candle lighted up his microscopic slide, the sight made Leeuwenhoek, with all of his experience, start. Some very little animals were "prettily amoving" through his witch's brew of a mixture. The biggest and longest shot through the drop like a pike (the fish). Another kind hovered together like a swarm of gnats. And if these specimens, from teeth as "clean and white as falleth to the lot of few men," were numerous and swift, they were as nothing compared to the myriad that inhabited a specimen Leeuwenhoek obtained from an old man who had never cleaned his teeth.

The human mouth was alive with animalcules, Leeuwenhoek reported to the Royal Society. He did not assume that the wriggling, somersaulting little animals that he found in the mouth and later in the wastes of the body were harmful. They were there, and Leeuwenhoek was satisfied to give explicit, unforgettable accounts and descriptions of them.

The Royal Society wanted to honor its prodigious correspondent. Leeuwenhoek was unanimously elected a fellow, and the society voted to send him his suitably inscribed diploma in a silver box. To increase its grandeur, the society voted at a later meeting to have its arms engraved on the box.

It was the greatest of honors. Leeuwenhoek was all but overwhelmed. He accepted, gratefully pledging the society that for the "singular favour" they had shown him, he would strive with all his might and main all through his life to make himself worthy of the honor and privilege. Universities might thereafter denounce him and the common folks might swear he was a conjuror who made people see what was not there, but from that time on, as a fellow of the Royal Society of London—not a mere foreign member—he was fortified against all assaults.

Word of the wonders the Dutch shopkeeper was disclosing spread everywhere, and Leeuwenhoek was besieged by visitors begging for a look at the fabulous little beasts. Peter the Great of Russia, on a visit to the Netherlands in 1698, sailed down the canal to Delft to see for himself. Two gentlemen of his retinue were dispatched to bid Leeuwenhoek to bring his incomparable magnifying glass to one of the Tsar's ships. The Tsar spent more than two hours peering through the Dutchman's revealing microscope and thus saw another of Leeuwenhoek's most famous sights.

In the tail of a little fish stuck into one of the tiny viewing tubes, the Dutch custodian showed his royal visitor the circulation of blood in the capillary vessels. By demonstrating how the blood went from the arteries to the veins, Leeuwenhoek confirmed the observation of the Italian scientist Marcello Malpighi and completed the theory of circulation developed by William Harvey some fifty-five years before.

Leeuwenhoek bountifully kept his promise to continue serving the Royal Society throughout his life. In all, more than two hundred of his inimitable letters went to London. As he lay dying in 1723, in his ninety-first year, his last act was to ask a friend to translate two last letters into Latin and send them to the society. Leeuwenhoek also had prepared one final gift for the great gentlemen and philosophers who had understood and valued his work.

It was a black-and-gilt cabinet filled with his most precious microscopes, some of which he had not previously displayed to anyone. In the five drawers of the cabinet were thirteen square tin cases, each covered with black leather and holding two instruments. A letter prepared in advance to go with them explained that every one of the lenses had been "ground by myself and mounted in silver, and furthermore set in silver, almost all of them in silver that I extracted from the ore, and separated from the gold wherewith it was charged; and wherewithal is writ down what object standeth before each little glass." Affixed to the needles were objects ranging from the globules of blood, "from which its redness proceedeth," to a thin slice of the wood of the lime tree, "where the vessels conveying the Sap are cut transversely," to "the organ of sight of a Flie."

More than two hundred years later, in the full perspective of history, Professor Lorande Woodruff of Yale called the Leeuwenhoek letters to the Royal Society the longest and most important series of communications that a scientific society has ever received.

Another part of the fantastically inhabited and structured universe that underlay and controlled the body and the surface had been revealed. It was another demonstration that men would have to look deeper, beyond the visible, for an understanding of what they are.

But when Leeuwenhoek died, there was no one to succeed him. His "secret" way of examining his helter-skelter of material had died with him, for he would reveal it to no one. Clifford Dobell, who made the modern translation of the Leeuwenhoek letters, believes that it was a method of darkground illumination. Either because of a lack of comparable techniques and instruments, or because of a feeling that there was no point in further pursuing studies done with such superlative thoroughness, no comparable microbe hunter appeared for the next 150 years.

In the fifty-nine years after Hooke had reported the odd little compartment in cork, cells had been found in all kinds of plant and animal tissue. The nucleus had been discovered, and Dutrochet had sensed and said that the cell was the "*pièce fondamentale*," the fundamental building block. And yet this growing knowledge of the cell was scattered. It had not been brought together with the completeness and definitiveness that would establish the cell as the universal unit of life.

This was still to come. It was awaiting the first meeting of two German scientists.

In 1837 they were introduced at a dinner party. Before the dinner ended, the two were so intent on their discussion that the rest of the company was forgotten.

The more voluble and assured of the professors, Matthias Jacob Schleiden, was expounding his views with a lawyer's force and persuasiveness—he had been a lawyer before he became a botanist. He, Schleiden explained, had been able to demonstrate that plants are composed entirely of cells and that their entire growth consists of a formation of cells within cells.

Dr. Theodor Schwann listened eagerly and put in many questions. Neither he, nor the great Johannes Müller, with whom he had studied, nor other anatomists had been able to find a single underlying elementary particle in animal tissue. On the contrary, the elementary particles of animal matter exhibited an overwhelming variety of form. It was true that cells could be seen under the microscope, but some were of one kind and some of another. In the nervous system the elementary particle was a fiber. And none of these various forms seemed to have anything in common except that they grew by the addition of new molecules between those already existing. As Schleiden knew, Schwann continued, the anatomists had never been able to discover any rule that governed the way molecules joined together to form living particles.

And yet, if there were a single principle, if animal tissues, like Schleiden's plant tissues, were made up of cells, it would explain some of the things he—Schwann—had been seeing under his microscope that very summer.

Schwann had been investigating the ends of nerves in the tails of frog larvae. As he put tiny bits of tissue under the microscope, he saw that the chorda dorsalis (spinal cord) had a "beautifully cellular structure." This confirmed Müller's finding that the chorda dorsalis in fish consists entirely of separate cells. Schwann at once asked if cells would be found in other tissues, and went on to examine some cartilage corpuscles. They too were made up of cells.

"But," he explained to Schleiden, "this led to no further results."

That was the state of affairs when he met Schleiden.

"Dr. Schleiden opportunely communicated to me his excellent researches upon the origin of new cells in plants and from the nuclei within the parent cells," said Schwann. "The previously enigmatical contents of the cells of the cartilages of the frog larvae thus became clear to me; I now recognized in them young cells provided with a nucleus."

Schleiden and Schwann hurried to Schwann's laboratory. The two scientists pored over Schwann's excellent drawings of the cell structure he had seen. Schleiden pointed out the similarity of the animal and plant cells. It was striking, and yet Schwann decided upon a thorough review and additional work before coming to any conclusions.

With the zeal of the explorer who sees his goal ahead, Schwann made new slides of the chorda dorsalis. The cells varied in size and had an irregular polyhedral shape, and each one had its own membrane. The cell walls were thin, colorless, smooth, and almost completely transparent. It was the same with the cartilage cells, no matter how closely they were examined.

"The detailed investigation of the chorda dorsalis led us to this result," Schwann explained. "The most important phenomena of their structure and development accord with the corresponding processes in plants.

"We have thrown down a grand barrier of separation between the animal and vegetable kingdoms, viz., diversity of structure."

The elation of the two scientists shone through Schwann's words, but Schwann knew that if the world was to be persuaded, complete proof would be needed. His work had already convinced him that the tissues had their beginnings in groups of young cells. But to establish the point, he decided, it would be necessary to prove it beyond question.

Schwann chose birds' egg as the best material to study—young, grayish-white eggs removed from the ovary. He put one on his dry object plate, pricked it with a needle, and allowed its contents to flow out. The contents consisted entirely of very pale cells of variable size. They were so small and so closely packed that they might have been taken for a granulous mass if they had not been viewed under a very revealing light. Schwann was able to show, however, that such cells subsequently become the globules of the yolk cavity.

He then took a fresh-laid hen's egg and removed a portion of the germinal membrane—a little white disk from which the embryo is formed. It was made up entirely of cells. Schwann demonstrated that after eight hours of incubation such cells gave rise to the first rudiments of the embryo.

The "primordial substance" itself was a cell, and from the moment of its origin to the formation of the tissues, development was only a multiplication of cells. In general, Schwann felt certain that he could say all tissues originate from cells.

But he knew this still would not be easily accepted. Nails, teeth, and bone derived from cells! Critics had always ridiculed the idea. To remove such doubts, Schwann decided that he would also have to trace such structures back to a cellular origin.

Schwann snipped a bit of nail from a finger of a newborn baby and divided it into longitudinal sections. The microscope showed that it was made up of layers of cells. Horny hoof tissue from the animal fetus also consisted entirely of "the most beautiful vegetable-like cells."

Feathers turned out to be another perfect example. It had always been maintained that they were "fibrous." To the naked eye they seemed to have nothing to do with the kind of cell that had been seen under the microscope for well over half a century. Schwann showed that in fact feathers are formed from large flat epithelial cells from the cortex. Each one of them had a "fine" nucleus and two nucleoli. The verdict was conclusive here too—fibers came from cells.

Teeth had traditionally been classified as bone. But when Schwann examined the crown of a growing tooth under the microscope, he found it likewise composed entirely of cells.

It was the same with the muscular fibers and nerves. Schwann could carry their generation back to cells and could show that the nerves "bring every part of the body into connection with the central part of the nervous system by means of uninterrupted cells." It was true, then, that even the least cellular-looking parts of the body originated in cells.

This work was completed in a little less than a year, and Schwann was ready in 1839 to publish his soon to be famous *Microscopical Researches into the Accordance in the Structure and Growth of Animals and Plants.*

He concluded it with a chapter on the "cell theory," thus naming and organizing one of the major concepts of biology and biochemistry. "All organisms and all their separate organs are composed of innumerable small particles of definite form," Schwann wrote. "There is one universal principle of development for the elementary parts of organisms however different, and this principle is the formation of cells."

Schwann had come a long way since he had been baffled by the cells in the tails of frog larvae.

There was another essential problem with which both Schleiden and Schwann struggled. Schleiden worked with it first. How did the cell divide and multiply?

Schleiden postulated that there were three places where new cells could be formed: on the surface of the previous mass of cells, in the interior of the cell, or in intercellular space. He did not have much difficulty in establishing which of his possibilities was the true one. "I believe that I have demonstrated that in accordance with Nature, the entire growth of plants consists only of formation of cells within cells," he said.

This was fine and true; time would unerringly verify it. Schleiden went wrong only when he tried to determine how the division occurred.

Schleiden held that the cell begins with a granulous mass, around which the nucleus forms as a sort of a coagulation. When the nucleus reaches its full growth, Schleiden said, a minute transparent vesicle arises on top of it, and this is the new cell. The botanist compared the new cell on the nucleus to a watch glass capping a watch.

Schwann more or less accepted this erroneous theory, or fancy, of his scientific colleague. Mitosis, or the process by which the cell divides to form two cells, was not to be discovered for many years. But Schleiden was not merely misinterpreting or failing to understand; he was inventing processes that did not occur.

Schleiden had mentioned his theory in a brief paper that he published in 1838, a year before Schwann brought out his book and reprinted Schleiden's essay at the end of it. Schleiden also emphasized that plants are built up of cells and that the embryo arises from a single cell, but his principal concern was with his theory of cell genesis, mistaken though it turned out to be.

Schleiden's error would soon be discovered and corrected, but his work and that of Schwann at last made the world aware of the cell. Perhaps their work only pulled together into a meaningful whole all the essential understandings that had been reached earlier by others; perhaps it only prompted a coalescing of knowledge. Certainly the world up to this time had not realized that the cell is the fundamental unit of living matter. Only after 1838 was it generally recognized that plants, animals, and men are compositions of cells, and that the life of the whole is a sum total of cellular life. It was an illuminating and heady concept, and one of the most fruitful ever opened to research. Men were on the way to understanding themselves.

In 1845 the Royal Society gave Schwann its highest honor, the Copley Medal. Two years later a translation of his work and Schleiden's was published in England by the Sydenham Society. In time, Schleiden and Schwann were cited in nearly all the textbooks as the founders of the cell theory. They were honored as the fathers of one of the most significant of all biological ideas, an idea that revolutionized a science.

The seemingly simple boxlike structure which Hooke first observed has revealed itself, under detailed examination by generations of scientists, to be a world in itself. The simplest organisms are born, react to their environment, reproduce, and die, all within the compass of a single unit. The most complex living things consist of billions of such units, differing among themselves according to the functions they perform. Inside their walls incredibly complex and not yet fully understood chemical reactions take place. The following chapter, an examination of these activities, is selected from the volume Biology, *written by three outstanding contemporary scientists. George Gaylord Simpson, Professor of Vertebrate Palaeontology at Harvard, is the author of such general books as* The Meaning of Evolution, Life of the Past, *and* Evolution and Geography *and has received numerous awards, particu-*

larly for his classic research on the evolution of the horse. Colin Stephenson Pittendrigh, formerly a faculty member of the University of Durham, England, and of Columbia University, has been associated with the Rockefeller Foundation and the Rockefeller Institute and is now Professor of Zoology at Princeton University. He has recently been active in investigating the possibility of life on other worlds. Lewis Hanford Tiffany has been on the teaching staffs of Ohio State University and Northwestern University, where he was William Deering Professor of Botany until 1962. He is an authority on plant physiology.

The Cell: CONSTITUENT MATERIALS AND PROCESSES

GEORGE GAYLORD SIMPSON,
COLIN S. PITTENDRIGH,
and LEWIS H. TIFFANY

EVERY CELL, every group of cells, every organism is dependent upon and influenced by other cells, other groups of cells, other organisms. Every organism lives in an environment of which it is itself a part, along with others of its kind, organisms of other kinds, and many nonliving substances such as air, water, and soil. No organism lives alone, and no living thing is sufficient unto itself. To consider a cell or organism apart from its environment is to consider the nonexistent, an abstraction that is not a real form of life. Relationships between life and environment are so pervasive a part of living that they are involved in all aspects of biology.

The fact that these relationships are so complex and that they do somehow affect every process of life makes it impossible to study them all at once or as a distinct and separate topic. It is necessary to take them up repeatedly, from different points of view and at different levels. Relationships to the environment here to be considered are mainly the exchanges and other interrelationships between cells and their environments: other cells and the various media (mainly fluids) in or on which cells live.

The degree of interdependence varies according to the properties of cells, their positions with respect to other cells, the kinds of organisms, and the surroundings of cells and organisms. One very essential sort of interdependence results from the fact that cells arise only from other cells. They are derived from one another in sequence through time. These aspects involve growth, reproduction, and heredity. Another sort of dependence is involved in the integration of cells in organisms. Some cells secrete hormones which modify the activities of other cells. Cells in organs and systems act to move or otherwise to affect the organism as a whole. Nerves (themselves cells) control and co-ordinate activities of other cells.

Perhaps the most fundamental interdependence of all is illustrated by the fact that all cells require certain sugars (and their derivatives) that are synthesized—or elaborated—in the first place only by the green cells of plants. Those green cells themselves are not independent. They require, at the least, supplies of raw materials from the environment and radiation from the distant sun. If they are parts of multicellular organisms they also require materials acquired or elaborated by other cells. In other words, every cell is dependent on sources outside itself for the materials of which it is composed and for the energy involved in its activities. Further, the rates of processes in the cell and also to some extent their nature depend largely on external influences such as light and heat. The present chapter will bear on these relationships: the sorts of materials and energy required by cells, the sources of these and how they get into the cell, and other effects of the environment.

The Materials of Life

One of the most fascinating things about life is the interplay of unity and diversity. From bacterium to sunflower and to man, all living things have much in common. Yet each sort of living thing is unique in its particular combinations of materials, processes, and relationships. All require just the same general sorts of materials. They require these in amounts and proportions which vary considerably, to be sure, but which vary within prescribed limits. Yet what they do with these materials is never precisely the same in any two sorts of organisms.

This unity-with-diversity can be strikingly illustrated by two exceptional cases. You would say—and, up to a point, quite correctly—that cellulose, the material of many plant cell walls and hence of wood, is a substance peculiar to plants as opposed to animals. Similarly, hemoglobin, the chemical that makes your blood red, seems completely characteristic of animals as opposed to plants. Yet cellulose occurs in a group of animals, the odd marine forms called tunicates, and hemoglobin has been found in the root nodules of some leguminous plants.

These exceptions show that the diversity among organisms is, after all, based on still more fundamental similarities. Cellulose is a complex compound built up from sugar molecules. Similar processes of building up sugar molecules occur widely in animals producing, for instance, the glycogen (animal starch) involved in muscular energy—and the point is further driven home by the fact that a few plants also make glycogen. Hemoglobin is a chemical combination of heme and globin. Compounds of the same general types are common in plants. For instance, chlorophyll, the green pigment of plants, is chemically very similar to heme. So the exceptional occurrence of cellulose in animals and of hemoglobin in plants illustrates a similarity in materials and in the way in which these are combined, even when the resulting combinations are different.

The material requirements of organisms are often summed up as food. Like many everyday words, "food" is an ambiguous word. No doubt bread and meat are food. But plain water is the largest material requirement of most organisms. Is water food? Vitamins are needed only in minute amounts. Are they food? Life requires relatively simple inorganic materials, such as water, used just so or as raw materials for more complex, organic compounds. The needed organic materials or foods are elaborated from raw materials by the organisms themselves, or are acquired by eating other organisms.

A LITTLE CHEMISTRY. It is necessary at this point to introduce some very simple chemistry, probably already familiar to you from general reading—and certainly less than you know if you have had an elementary course in chemistry or physical science. You surely know that the smallest particles in nature, electrons, protons, and the rest, are commonly organized into *atoms.* Atoms represent in most basic form the chemical *elements,* the building blocks for all larger chemical units. In view of the tremendous complexity of the matter composed of them, the number of elements is surprisingly small. Only 101 are known; several of them have been made artificially and may not occur in nature. About 35 are common in nature and important for life.

You also doubtless know that two or more atoms (up to thousands) can combine with each other, linked together by electrical forces, to form a *molecule.* If the molecule contains more than one kind of atom, it is a *compound.* Atoms and molecules or both can *react* with each other and produce different kinds of molecules. In ordinary chemical reactions, the number and kind of atoms are the same before and after the reaction. Only their combinations in molecules change.

Each element has a name and an abbreviation that stands for the name. The elements most important in biology are: hydrogen, abbreviation H; oxygen, O; carbon, C; and nitrogen, N. The simple (or empirical) formula for a molecule shows what atoms it contains. Water is H_2O, a formula which means that it has two atoms of hydrogen to one of oxygen. Reactions are written in terms of such formulas. The reaction for forming water from its elements is:

$$2H_2 + O_2 \rightarrow 2H_2O.$$

That means that two molecules of hydrogen (each with two atoms) combined with one of oxygen (also with two atoms) produced two molecules of water. There were four atoms of hydrogen and two of oxygen to begin with, and of course the same six atoms are there in the water after the reaction. One other point is that there is less energy in two molecules of water than in two of hydrogen and one of oxygen. In this reaction, energy is released from chemical form and appears in some other form such as heat. The most meaningful and complete formula for the reaction is therefore:

$$2H_2 + O_2 \rightarrow 2H_2O + \text{Energy (heat)}.$$

Some other simple chemical principles will be mentioned later, but this is all you need to know for the time being, and in fact this is the whole basis of chemistry. (It is true that things a bit more complicated are built upon this simple basis!)

INORGANIC MATERIALS. All the materials of life are *ultimately* derived from relatively simple inorganic[1] compounds and elements. Once they are assimilated by a living thing, these ultimate materials often become extremely elaborated and are passed on from one organism to another. Most of these elaborated chemicals are finally broken down into simple, inorganic forms again. There is, then, a cycle, with the inorganic materials at the beginning and end; it thus seems logical to begin consideration of life's materials with them. They include water, carbon dioxide, oxygen, nitrogen, and a great varicty of what may be called, as a group, mineral salts.

Water. Protoplasm is largely composed of water—plain water, H_2O. Higher plants and animals also contain much fluid that is not in the protoplasm but which bathes the cells or moves among them in vessels. This fluid, exemplified by the sap of trees or the blood of animals, is essentially

1. "Organic" chemical compounds are complex molecules, all of which contain carbon and occur in nature only as the products of living organisms, or, nowadays, as the product of man's ingenuity in a chemical laboratory. Inorganic compounds occur in nature independently of the living organism.

water with a complex of inorganic and organic materials dissolved in it. The liquid part (plasma) of blood is about 90 per cent water; most active protoplasm contains about the same percentage of water. We are all familiar with the discomfort of animals (such as humans) and the wilting of plants when deprived of water. Some dormant structures, such as spores and seeds, can retain sufficient water to remain alive for months or even years without new supplies. Nevertheless, all organisms eventually die if their water supply continues to be inadequate.

Water plays three main roles in living things: (1) it is a medium in which other materials move from place to place; (2) it is the seat and facilitator of chemical reactions; (3) and it enters into reactions itself, and goes to make up some of the more complex materials of life. So many other substances dissolve in water that it is sometimes called a universal solvent. It is true that some of the materials necessary for life are scarcely or not at all soluble in water, but even these are commonly surrounded by water or suspended in it as colloids or emulsions. Water is probably involved in *all* the chemical reactions in cells, either as a solvent or as a reagent.

Some water used by life arises within the cells as a result of chemical reactions. For instance, the complete oxidation of fats produces large quantities of water. Some desert animals never drink liquid water, obtaining enough from metabolism of fats (also of carbohydrates). Generally, however, water as such must enter the organism from its environment, getting into individual cells from outside them. Water is also constantly lost from cells and organisms. The necessary balance between intake and outgo of water involves some of the most interesting processes and problems of life.

Water, as such, is thus a large and necessary part of all organisms. The role of water as a raw material for other compounds is no less essential. Practically all organic compounds contain hydrogen, the ultimate inorganic source of which is for the most part water, H_2O. Some other inorganic materials taken in by living things supply hydrogen in chemical combination, but this constitutes only a small fraction of what the organism uses. Hydrogen as an element, H_2, occurs in nature as a gas, but it is not abundant and is never utilized by living things. Water also supplies much of the oxygen in organic compounds. Oxygen, like hydrogen, occurs in nature as a gas; but unlike hydrogen, this elementary source is exploited by living things.

Carbon dioxide. It is the definition of *organic* compounds that they all contain carbon. Once it is incorporated in a living thing, a carbon atom may pass through hundreds or thousands of different combinations within one organism and then, commonly, within others and yet still others as the materials are passed on. The beginning and end of the cycle in the inorganic environment involve carbon in the simple compound CO_2, carbon dioxide. Under ordinary conditions this is a gas. (At temperatures lower than those in nature it solidifies; solid CO_2 is what we call dry ice.) The gas forms about 0.03 per cent of the atmosphere. This mere three-hundredths of one per cent is the main inorganic reservoir and source of carbon as a material for life.

The gas CO_2 does not usually enter directly into chemical reactions. However, it dissolves in water (as everyone knows who has ever opened a bottle of carbonated drink) and then reacts readily in various ways. This activity is due in part to the fact that CO_2 in solution reacts with water, itself and forms a weak acid, carbonic acid, H_2CO_3, in accordance with the following equation:

$$CO_2 + H_2O \leftrightarrows H_2CO_3.$$

The double arrow, $\leftrightarrows$, indicates that the equation, like a great many of those important in life processes, can readily go in either direction. Slight changes in pressure, heat, concentration, or presence of other chemicals can reverse the reaction:

$$CO_2 + H_2O \rightarrow H_2CO_3,$$

and produce the opposite:

$$H_2CO_3 \rightarrow CO_3 + H_2O.$$

The beginning of the incorporation of atmospheric CO_2 into the materials of life is a more complex series of reactions with water. In green plants these two raw materials are combined into simple sugars. This is an extremely important and basic synthesis for the whole world of life. From the simple sugars, the carbon is passed on into many other substances.

The end of carbon's participation in the chemistry of life usually involves its withdrawal from more complex compounds and its combination with oxygen. It thus forms CO_2 again, most of which finds its way more or less directly back into the atmosphere. Almost all organisms, both plants and animals, are constantly forming CO_2. The accumulation of this gas, or of H_2CO_3 readily formed from it, in cells or in fluids such as blood soon becomes harmful. Its elimination is therefore necessary.

Oxygen. The great majority of the compounds involved in the substances and processes of life contain carbon, hydrogen, and oxygen. Some contain no other elements. Others contain quantities, usually much smaller quantities, of various additional elements. The inorganic source of much oxygen in organic compounds is water. In addition to water and oxygen-containing organic foods, most organisms also require extra oxygen in the form of the element itself, O (or, as a molecule, O_2). Along with water and some salts, this is an inorganic material that can be utilized *directly* by animals as well as plants, in marked contrast to the carbon source, CO_2. (There are, however, some lowly organisms, mainly bacteria and parasites, that can live without oxygen and may even be killed by it.)

Oxygen is a gas, and its great inorganic reservoir is the atmosphere, which contains about 20 per cent (by volume) of elemental oxygen (that is, not combined with other elements). Oxygen dissolves readily in water, remaining in the form of the element, and so is also available to aquatic organisms. In fact, oxygen must be in solution in water to enter organisms living in air.

The principal role of elemental oxygen in cells is to combine with carbon and hydrogen, from the breakdown of organic compounds, producing carbon dioxide and water. The water so formed may be further utilized by the organism, but the carbon dioxide is almost entirely eliminated. The input of oxygen and output of carbon dioxide depend on the chemical activity of the cells and of the organism as a whole. Oxygen consumption is a fairly good measure of total metabolism in most organisms. If you have ever taken a basal metabolism test, this was the principle involved.

If carbon (for instance, charcoal) is burned in air, it combines in a simple way with oxygen and produces heat, which is, of course, one form of energy:

$$C + O_2 \rightarrow CO_2 + \text{Energy.}$$

Because of this reaction, so familiar to everyone, and because the body's oxygen consumption and carbon dioxide elimination do tend to be proportional to total activity, it used to be supposed that some such reaction accounts for the body's heat and other energy. The notion is held by many people that "fuel"—organic compounds such as sugars or fats—is "burned"—combined directly with oxygen—in the body and that this is how we obtain energy. It is now known, however, that this is rarely if ever a significant source of useful organic energy. Most such energy is released by chemical reactions that do *not* use oxygen. As a rule, oxygen is used and CO_2 is formed only *after* the important energy release, in a sort of sweeping up of the debris of earlier reactions and setting up for their repetition.

Animal metabolism utilizes but does not produce elemental oxygen. The process of combining CO_2 and H_2O to form sugars in green plants, however, involves release of O_2, and while this synthesis is going on it usually produces more oxygen than the plants need. Green plants in light thus usually give off oxygen to the surrounding water or air. In the dark, when the active phase of sugar production is not proceeding, plants produce no spare oxygen but continue to utilize oxygen and give off CO_2. This is why fish in a pond with green plants may be suffocated during the night. The dissolved oxygen in the water is being used up and none is being produced by the plants.

Nitrogen sources. After carbon, hydrogen, and oxygen, the most common element in protoplasm and in the materials of life is nitrogen, N (in molecules, N_2). It is, in particular, a constituent of all proteins, an extremely important class of organic compounds to be discussed later in this chapter.

There is a tremendous store of nitrogen in the atmosphere, which is (by volume) almost 80 per cent elemental nitrogen. Yet no animals and few plants can make direct use of this nitrogen. Most plants can utilize nitrogen from the environment only if it is in the form of various inorganic compounds: ammonia (NH_3) or its compounds; nitrates (salts containing NO_3); or nitrites (salts with NO_2). Animals need to take on nitrogen in the form of compounds, especially proteins, which they obtain from plants or from other animals. The withdrawal of nitrogen from the atmospheric reservoir and its incorporation into life thus depend on processes that make ammonia, nitrates, and nitrites. Some inorganic processes do this, especially lightning, but formation of these compounds is due more largely to a few kinds of organisms that are able to metabolize elemental nitrogen—to fix it, as it is said. Some bacteria, and simple plants (some algae, and fungi), can do this. Most noteworthy are bacteria that live in nodules in the roots of beans and related plants (legumes).

On the other hand, few organisms decompose compounds of nitrogen completely and produce the element nitrogen. Some denitrifying bacteria do this, but the end products of nitrogen metabolism in most organisms are still organic compounds, such as urea or uric acid in animals, or, at the simplest, ammonia. Since even ammonia can be utilized by some plants and is readily turned into nitrates and nitrites by certain bacteria, the nitrogen remains available to life for long periods of time and passage through many different organisms.

Mineral salts.[2] Besides C, H, O, and N, many other elements are required in smaller amounts by all forms of life. The inorganic sources of these are, in most cases, salts dissolved in water—water in the soil or in lakes, streams, and seas. Plants and, to some degree, aquatic animals usually acquire these salts directly from the water of the environment. Land animals also acquire and utilize salts directly to some extent, but obtain much of their mineral requirement more or less incidentally along with their organic food. This is one reason for emphasis on a balanced diet by physicians—we have not only the obvious requirements for building materials and energy in our food, but also the need for small amounts of many different minerals, not all of which are likely to be present in the required forms and quantities in any one food. In exceptional cases extra quantities of iron, calcium, iodine, or sodium may be needed and supplied medicinally, or reduction of mineral intake (for instance, of table salt) may be indicated.

Among the mineral elements, sulfur, phosphorus, potassium, sodium, calcium, magnesium, chlorine, iron, and copper are required by most or all organisms, although plants do not need sodium. Other elements are widely required, but perhaps not by all forms of life. For instance, in addition to those listed, iodine, manganese, zinc, and fluorine are elements essential in minute amounts to man and numerous other organisms. It has been found that elements present in such small quantities as to have escaped earlier detection may nevertheless be absolutely necessary to some organisms. For example, the mysterious wasting away of sheep and cattle on some pastures was found to be due to a lack of cobalt. At present a great deal of research is being done on these "trace elements." The need for extremely small traces of a great many elements is turning out to be much more widespread than had been realized.

ORGANIC MATERIALS. Organisms vary greatly in their abil-

2. Strictly speaking, the main inorganic nitrogen sources for life are actually mineral salts, and therefore belong under this heading, but they have such a special role that they have been briefly treated separately in the preceding section.

ity to build up the compounds needed for their lives. Plants generally can make all the organic materials they require, and animals generally cannot. This is a matter of nutritional needs and processes. Here we must note that various organic substances are essential *materials* for life and must somehow be acquired by all cells. This is true whether the cells make these materials for themselves, receive them from other cells of the same organism, or ultimately derive them from sources quite outside the organism. The materials to be discussed at this point fall into the very broad chemical groups of carbohydrates, fats, and proteins. There are other groups just as essential to life, notably the vitamins and hormones. They are, however, more clearly understood in connection with special processes considered later.

Carbohydrates, fats, and proteins include the principal building materials of cells; they are also the main sources of energy. These roles are not always, or even usually, clear-cut in the cell. It is quite common for the same material to act successively in building and in energy production. It may, indeed, do both at the same time. Proteins may be structural elements, sources of energy, and facilitators of reactions (catalysts) all at once. In the cell there is a constant, intricate, interweaving flow of materials and processes. The processes are not like building a house and then hauling in fuel and making a fire in the stove. The cell house is self-energizing; its materials are always in a state of flux; the general plan persists, but every brick or board is continuously being modified or torn out and replaced.

Carbohydrates. Carbohydrates have a particularly basic role as materials for life, for through them inorganic carbon is first incorporated into living things. Carbohydrates also contribute to the formation of other sorts of organic compounds. Carbohydrates are important building materials, and they are especially involved in the chief energy-releasing processes in both plants and animals. They are composed entirely of carbon, hydrogen, and oxygen. Their molecules always contain just twice as many atoms of hydrogen as of oxygen—the same proportions as in water. Common examples of carbohydrates are sugar, starch, and cellulose.

Most carbohydrate molecules have a basic unit of six carbon atoms linked together in various ways, with hydrogen and oxygen atoms attached around them in different ways. Because of the differences in attachments, two carbohydrates may contain the same numbers and kinds of atoms and still be quite different substances. Both fructose (fruit sugar) and glucose (a principal ingredient in corn syrup) have the formula $C_6H_{12}O_6$. This sort of formula, called *empirical,* shows the atoms present but not their arrangement. The arrangement in fructose and glucose differs in a way that makes them two distinct sorts of sugars, as we shall see.

By reactions commonly occurring in life processes, two or more groups of six carbon molecules of the simplest carbohydrates (simple sugars) can be bound together. When this happens between any two such molecules, two atoms of hydrogen and one of oxygen are dropped out of the compound. Those three atoms lose their places of attachment in the carbohydrate molecule and appear combined with each other, as water:

$$\underset{\text{Glucose}}{C_6H_{12}O_6} + \underset{\text{Fructose}}{C_6H_{12}O_6} \longrightarrow \underset{\text{Sucrose (table sugar)}}{C_{12}H_{22}O_{11}} + \underset{\text{Water}}{H_2O}.$$

(The process as it really goes forward is more complex. The equation sums up only what goes into the complex process and what comes out of it.)

Essentially the same process can link together large numbers of the six-carbon units and attached hydrogen and oxygen atoms, eliminating one molecule of water for each unit added:

$$\underset{\substack{n\text{ glucose molecules}\\ \text{added together}}}{n\,(C_6H_{12}O_6)} \xrightarrow[\text{or dehydration synthesis}]{\text{condensation}} \underset{\text{Starch}}{(C_6H_{10}O_5)_n} + \underset{\substack{n\text{ molecules}\\ \text{of water}}}{n - 1\ H_2O}.$$

(Of course, this process as it actually occurs in cells goes through many steps and is even more complicated than the last.)

Here n indicates the number of six-carbon units involved, a large number in this case. The result is a relatively large, complex, long molecule, called a *polysaccharide;* the term rather neatly sums up what has happened: it means "many sugars." This building up of a large molecule from simple sugar units is called *condensation;* it is a dehydration synthesis. It is one important way in which the materials of life are compounded. Like simple sugars, different polysaccharides may have the same empirical formula and yet be quite different because the atoms are arranged differently. For instance, cellulose and glycogen (an energy source in muscle, mentioned earlier) have the same general formula as starch:

$$(C_6H_{10}O_5)\ n.$$

Starch and glycogen are ready forms of storage of carbohydrates, starch usually in plants and glycogen usually in animals. They are not soluble in water and so can remain in cells unchanged until required. A potato is mainly a store of starch, and in animals glycogen is stored in the liver, among other places. When the carbohydrate is utilized, it becomes soluble by the reverse of the summary equation given above:

$$\underset{\text{Starch or glycogen}}{(C_6H_{10}O_5)\ n} + \underset{n-1\text{ molecules of water}}{n - 1\ H_2O} \xrightarrow{\text{hydrolysis}} \underset{n\text{ molecules of (soluble) glucose}}{n\ (C_6H_{12}O_6)}.$$

You see that we could have written the equation $\leftrightarrows$ showing that, like all other chemical processes in life, this can and does go both ways. This reverse of condensation is called *hydrolysis*. It is an essential part of the digestion of food, so much so that it is sometimes called "digestion." It is, however, also involved in processes not normally thought of as digestion, such as the release of blood sugar from the liver.[3]

Fats. Fats are also composed of carbon, hydrogen, and oxygen, but the ratio of hydrogen to oxygen is much greater than two to one, the ratio in carbohydrates. For instance, the empirical formula of stearin, a fat, is $C_{57}H_{110}O_6$. The way in which the molecules are put together is also characteristic. They resemble polysaccharides in being formed by the condensation of simpler units, but those units (fatty acids and glycerols in chemical terms) are markedly different from simple sugars. A fat can also be hydrolyzed back to its constituent units.

Fats are insoluble in water and are often stored in both plants and animals. They have other essential roles, but fats are a rich energy source. Their complete metabolism entails release of over twice as much energy as for the same weight of carbohydrate. It also involves the formation of much more water, an important point for organisms in environments from which it is difficult to obtain sufficient water in a more direct way.

Proteins. The proteins are the stars of the biochemical opera. They include the largest and most complicated of all known molecules. In a sense, it is silly to say that one substance or another is more important when all are necessary. This is true of life's materials, from common water to the biggest protein. Yet proteins do have a very special importance. The complexity and diversity of life itself is largely dependent on the complexity and diversity of proteins. The proteins can, indeed, be said to direct the whole show in addition to starring in it, for the chromosomes, themselves partly protein, exert their control over the life of the cell through the agency of specific proteins (enzymes) they manufacture. Thus proteins determine what will develop from a seed or an egg. From that point of view, we are what

3. Since we have said that starch is insoluble, has it occurred to you to wonder why a starchy soda cracker dissolves if you chew it long enough or even just hold it in your mouth? The answer is that saliva contains a substance that hydrolyzes starch.

proteins made us. Even after development, proteins largely determine how the organism operates. Of course, they do so in conjunction with other materials, but theirs is the essential direction of most of the processes. They are also structural materials and sources of energy. Perhaps, after all, it is not out of line to say that proteins are among the two or three most important kinds of chemicals in the universe!

The large molecules of proteins are built up of smaller units in much the same way as are polysaccharides and fats. That is, the process is one of *condensation* (dehydration synthesis), with loss of a water molecule for each unit added. The units are, of course, quite different from those of polysaccharides and fats. In the case of proteins, the units are the *amino acids,* which have the general formula $CHR\ (NH_2)\ COOH$. In this formula C, H, N, and O stand for carbon, hydrogen, nitrogen, and oxygen in the usual way. R stands for other elements in the molecular structure. In the simplest amino acid, glycine, R is just a hydrogen atom, so that glycine is $CH_2\ (NH_2)\ COOH$. In other cases R is more complex, sometimes more complex than the central amino acid group to which it is bound.

The number of amino acids cannot be precisely stated at present. Twenty-four are reasonably established as distinct, of known formula, and present in the proteins of living organisms. All proteins contain many amino acid units. Ten thousand or so is a usual figure, and some proteins may contain a hundred thousand or more. The number of possible different arrangements of twenty-odd kinds of units in groups of ten to a hundred thousand is more than astronomical—it is completely inconceivable. It is much more than the total number of particles in the visible universe! There is, however, reason to think that not all the mathematically possible groupings are chemically or biologically possible. Certain sequences of amino acids apparently tend to recur in a fixed way in proteins. This reduces the possible number of different proteins, but still leaves an immensely large figure.

Only plants[4] can make *all* the amino acids they need and

4. Some fungi are probably exceptions.

later fabricate them into complex proteins. Animals always depend on plants for some amino acids. Man, for instance, has to receive in his food nine of the twenty-four amino acids he eventually utilizes in protein synthesis; the other fifteen he manufactures from the basic nine. All organisms, without exception, manufacture proteins from amino acids, and it is not surprising that sensitive tests reveal that they all make some proteins peculiar to themselves. There must be, therefore, well over a million different kinds of proteins in the whole world of life, and the number may be much greater. Of these million or more proteins, only some five hundred have, however, been isolated and identified by biochemists.

Besides the proteins, strictly speaking, there are *conjugated proteins.* These have, in addition to the amino acids, nonprotein groups chemically bound onto the molecule. The addition is called a *prosthetic group*. Two examples will give some idea of how important conjugated proteins and their prosthetic groups are. The directive activity of proteins in heredity, development, and other processes is particularly that of the nucleoproteins, conjugated proteins with a prosthetic group containing nucleic acid. Hemoglobin, on which our cells depend for oxygen, is a conjugated protein, globin, with a prosthetic group called heme.

The empirical formulas of proteins do not mean much; their amino acid composition is much more significant. Here are, however, the formulas of a few rather simple proteins, just to suggest how large the molecules are in comparison with most molecules: casein of milk, $C_{708}H_{1130}O_{224}N_{180}S_4P_4$; gliadin of wheat, $C_{685}H_{1068}O_{211}N_{196}S_5$; human hemoglobin, $C_{3032}H_{4816}O_{872}N_{780}S_8Fe_4$. (S is sulfur, P phosphorus, and Fe iron; you know the other symbols.)

Most proteins are soluble in water. Their molecules are, however, so large that a solution of proteins does not act like an ordinary solution of small molecules. It acts more as if the molecules were large clumps of molecules of more usual size. In other words, a protein solution is more like a colloid.

The sizes of various molecules have a decided bearing on the different processes of life. A small molecule, for instance, may pass through a cell membrane readily when a large molecule cannot do so at all. This brings us to some consideration of how the materials of life are obtained, how they get to where they are needed, and how they move into and out of cells through the cell membranes.

How Cells Get Their Materials

In every living cell some particular compounds involved in its activities are made within the cell, and others come from outside. Cells vary enormously as to what compounds they make and what they take in, and also what they release. In multicellular organisms there is great diversity in this respect among the cells of one organism. Their cells are chemical specialists. There is also great diversity among different organisms. We have already seen that green plants can make sugars and most of the other essential foods, while animals cannot. These are matters of nutrition and of food chains among organisms in communities, above the level of cells. At this point, we are more concerned with the fact that each cell must obtain materials from outside itself, regardless of what compounds the cell itself makes or what it does with the materials taken in. The materials must also move within the cell.

Cells usually obtain their materials as molecules or parts of molecules (especially the *ions* of the chemists) from watery solutions. There are some real and some merely apparent exceptions to this generalization. Cells in direct contact with air may acquire and lose gases without there being an external liquid solution. (The gases are usually in solution when *within* the cell.) Even in cases apparently of this sort an external solution may really be involved. Your lungs do not extract oxygen *directly* from air, but from solution in a thin liquid film that covers the cells at the surface within the lungs. There are also cells that take in undissolved material from their surroundings, small bits of solid food, globules of fat, or the like. You may see an amoeba surrounding a food particle and taking it whole into the cell. There are cells in your own body that do the same sort of thing. Among them are certain cells (phagocytes) in your blood that surround and digest bacteria, an important part of resistance to disease. This is more of an apparent exception than a real one. The particles taken in whole by an amoeba or other cell are reduced to molecules in solution before they actually enter the protoplasm and take part in its chemical activities.

Thus even in these cases the rule holds that protoplasm usually obtains materials from a solution with which it is in contact. A cell, or rather a protist, living alone obtains its materials directly from the surrounding inorganic environment. The environment of such an organism is usually water and therefore a solution. Chemically pure water does not occur in nature, and no organisms can survive indefinitely in really pure water. All water in which life exists is a weaker or stronger solution. Cells in multicellular organisms are usually also in constant contact with solutions, solutions in adjacent cells or, commonly, in liquids moving outside but among the cells. These liquids, of which the sap of a tree or your blood plasma are special examples, are generally very elaborate solutions, intricately compounded and influenced by the chemistry of the whole organism.

The question of how cells get their materials thus reduces to the question of how molecules move about, sometimes in gases, but more often in the liquids in which the molecules are in solution.

HOW MOLECULES MOVE ABOUT. Some movements of molecules are sufficiently obvious to require little discussion. If a molecule is in solution in a liquid or is part of a gas, it is moved along by movements of the liquid or the gas. As usual, there are many complications and a long string of "why's" back of this fact, but the simple fact may suffice us here. It is a less obvious fact that molecules also move about under their own power, so to speak. This movement involves several principles, among which the most important for present purposes is that on an average more molecules move away from than toward a region where such molecules are especially numerous. This is the principle of *diffusion*. Molecules or atoms also move through a membrane. Principles here bear on the fact that some molecules or atoms move more easily than others through certain membranes and that some sorts of molecules or atoms under given conditions may move through a membrane in greater number in one direction than in the other. These are the principles of *semipermeability* and of *osmosis*.

The importance of principles involving membranes is evident from the fact that cells are enclosed in membranes. Materials have to pass through the cell membrane and therefore are subject to the principles of semipermeability and osmosis. These processes and also diffusion (closely related to them) are going on all the time throughout our bodies and in active parts of all other living organisms. They are so fundamental for the properties and activities of life that we must review them here.

Molecular movement. It is a physical property of matter, even of solids, that molecules are always moving. A simple demonstration of this is to uncork a bottle of perfume in a closed, still room. You may stand several feet from the bottle, but soon you will smell the perfume. Molecules have moved out from the bottle and through the air of the room. Your nose, which is capable of some of the most delicate chemical tests known, has detected the dispersed molecules.

Molecular movement in gases, such as that of the perfume molecules in air, is considerably faster than in liquids.[5] It is faster in liquids than in solids, although it also goes on among all the molecules of the densest solid. It is faster in hot than in cold substances; in fact, the movement of molecules *is* the phenomenon we call heat.

Molecules are much too small to see. Consequently we cannot actually or directly see molecular motion, but there are simple ways to see it indirectly, by some of its visible results. Put a little face powder in a drop of water and look at it under the high power of an ordinary microscope. The particles of face powder (which are composed of many

5. The relatively great speed with which perfume molecules reach our nose is due only in part to the greater speed of diffusion in gas than liquid. In any normal room there are air currents that speed their passage. Currents or turbulence in water will also speed the movement of molecules in it.

molecules) will be seen dancing about. They are being bombarded by the moving water molecules and are small enough to move when hit by a molecule.

Diffusion. In a drop of water millions of molecules move about virtually at random. The *average* result, the net effort of these movements in every direction, is nil. In spite of all this activity within it, the drop does not go anywhere, nor do the molecules become more concentrated in one part of the drop than in another. This is the usual situation when molecules are evenly distributed through the space or substance being studied, even where there is a uniform mixture of different kinds of molecules.

When you opened the perfume bottle, however, something else happened. There was a change in net effect; the perfume molecules did go somewhere. When the bottle was opened, there were no perfume molecules outside it. Soon there were such molecules at increasing distances away from the bottle. A short time later the molecules were still *highly concentrated* in the bottle, but also were highly concentrated near it. Concentration away from the bottle was progressively *lower,* or was zero before perfume reached a given part of the room. The motion of each molecule was random. Some even moved back into the bottle, but *more* moved out. The net effect, the average over millions of molecules, was that more molecules moved from regions of high to regions of low concentration.

This tendency for molecules to spread from regions of higher to those of lower concentration is quite general and is called *diffusion.* If it continues, the concentration eventually becomes the same everywhere. The perfume molecules become evenly distributed throughout the room. There is a state of *equilibrium.* What happens then? Does diffusion stop? Do the perfume molecules stop moving?

Diffusion of molecules in a gas, quite analogous to our example, takes place in some life processes. In many land plants, leaves have gas-filled spaces between the cells, spaces that have special openings to the outer air called "stomates." When the cells take up CO_2 from the air spaces, the concentration is lowered there and more CO_2 diffuses in through the stomates from the outside atmosphere. When the cells give up O_2 into the spaces, its concentration is there increased and O_2 diffuses out into the atmosphere.

Now consider another simple experiment. If a crystal of copper sulfate (or any soluble, colored compound) is put in the bottom of a glass and the glass is filled with clear water, the water soon begins to turn colored around the crystal. The zone of colored water becomes larger from day to day. What is happening? At the surface of the crystal, copper sulfate is going, molecule by molecule, into *solution* in the water, the *solvent*. Naturally the dissolved molecules are more concentrated right at the surface of the crystal than elsewhere. They therefore diffuse into the surrounding water, and so the colored zone spreads.

It spreads very slowly, however, much more slowly than the spread of perfume molecules in air. The perfume can be smelled three feet from the bottle within a few minutes, but it takes more than a year for copper sulfate to diffuse this far through water in recognizable amounts. If you have ever tried to run in a couple of feet of water, you know the reason for this difference in rate of diffusion. Water is a *denser* medium than air. Down at the level of the molecules, this means that in a gas the molecules are farther apart. Molecules are less likely to collide, and the diffusing molecules are less often slowed up or bounced back. In a liquid the molecules are much closer together and the diffusing molecules are slowed up much more. In either air or water, however, the molecules are very small in proportion to the space between them. There is plenty of room between them for the oncoming molecules of copper sulfate, and the total volume of air or water is not increased by the diffusion.

In those examples, then, diffusion did not push the molecules of air or water farther apart. On the other hand, in a solid or a dense colloid the molecules are so close together that diffusion among them may force them apart. If a thin piece of dry gelatin is placed in water, it swells. Water has diffused into the gelatin and pushed its molecules apart. The resulting volume is greater than that of the dry gelatin alone but less than that of the dry gelatin plus the *original* volume of the water diffused into it. This particular sort of diffusion is called *imbibition.* The word simply means "drinking in," but is perhaps a more elegant way to put the matter.

Imbibition is a familiar part of daily life and of life processes. It makes wooden doors stick in rainy weather, and it makes bean seeds swell and burst their seed coats just before they sprout. The pressures caused by imbibition may be enormous. Imbibition in starch can develop pressures up to fifteen tons per square inch. Ships loaded with rice have been split apart when water got into the holds and was imbibited in the rice.

Semipermeability. Now that we have discussed diffusion in general, we can take up diffusion through membranes, a special circumstance of peculiar importance in the life of cells. The first point is that some substances diffuse quite readily through some membranes. The membrane is no particular barrier to the given substance and is said to be *permeable* to it—a fairly obvious and clear special application of an everyday word. Of course, if a substance cannot pass through a membrane, the membrane is impermeable to it. If a membrane is permeable to some substances and less permeable or impermeable to others, it is called *semipermeable* (or "differentially permeable," which may be more precise but is clumsier).

The point of this for our study of life is that cell membranes are semipermeable. They are generally permeable to water and impermeable to colloids. The membrane holds in the colloidal elements of the cytoplasm and yet permits water to move rather freely into and out of the cell, an essential feature of life processes at the level of cells. The cell membrane is also more or less permeable to various materials dissolved in water. Thus needed materials from outside can diffuse through the membrane. Once within the cell, they can be held in the colloidal mesh or in compounds to which the membrane is not permeable. Waste products can be converted into forms to which the membrane is permeable and so can leave the cell.

Actually the situation is more complicated than that, and in a very interesting way. Like other parts of cells, their membranes are in a continual state of flux. The degree of permeability to various substances is not constant, varying quite markedly from time to time. Thus even without any change in its own state or composition, a substance may pass through the membrane at some time and be held back by it at others. The variation depends on many things, such as sugar content, acidity, electrical properties, or conditions of colloids within the cell, and outside the cell, temperature, acidity, heat, light, concentration of materials, and other factors. Alterations in membrane permeability greatly influence the kinds and amounts of materials diffusing into and out of cells. This is, therefore, one of the essential mechanisms in the regulation of basic life processes in the cell.

We have noted that cells need to obtain and to lose certain materials at various times. The membrane's permeability corresponds with these needs and is a mechanism that helps, when all goes normally, to ensure that the needs are met. You might be tempted to say that the membrane has these characteristics *because* of the cell's needs—and there you would part company with science. That sort of answer slips in a hidden metaphysical postulate that needs can *cause* their fulfillment; it is therefore to be rejected alike by science and by straightforward common sense. If you are at all inclined to balk at the conclusion consider facts of the following sort. Cell membranes may pass materials of no possible use to the cell (such as nitrogen into green plant cells); they may let out so much water that the cell dries up and dies; they may let in poisons that kill the cell.

Osmosis. Try another experiment. Make a container of a membrane permeable to water: a pig's bladder or a frog skin is a natural membrane of this sort, or a collodion sack can be prepared. Fill it with water, close it tightly, and immerse it in water. Nothing noticeable will happen. Although the container is permeable to water, the concentration of water is the same inside as outside. Therefore as many molecules move out as move in, and equilibrium or the *status quo* is maintained. Now put a sugar solution in the container (preferably leaving it somewhat slack but without air inside) and immerse it in water (preferably distilled water) again. The container will swell up and become turgid.

What has happened? Evidently more water has moved in than out through the membranes. Why? What you already know of diffusion supplies the answer. Water molecules are less concentrated in the sugar solution than in pure water. Molecules move predominantly from the pure water into the solution. If the membrane were equally permeable to sugar and to water, sugar molecules would also move out of the container. But the membrane is semipermeable, being much more permeable to water than to sugar. So no (or very little) sugar moves out of the container, more water moves in than out, the amount of fluid in the container increases, and the container swells. That is the process of *osmosis.*

The influx of water into the container produces pressure, *osmotic pressure.* The flow through the membrane continues until the pressure inside forces water molecules out through the membrane as fast as osmosis brings them in. The pressure then ceases to rise, becoming steady. A state of equilibrium has been reached. The amount of osmotic pressure that can develop in a solution separated from distilled water by a semipermeable membrane depends on concentration of the solution, temperature, and other factors. Under given conditions, the pressure has a characteristic, constant value for any soluble substance. This possible pressure under standardized conditions is the *osmotic value* of the substance. Usually that value is not expressed directly in terms of pressure but by some correlated figure. You will often find it expressed in terms of a change, symbolized as Δ, in freezing point of a solution as compared with pure water. The relationship is rather complicated. But it may be convenient for you to know that $\Delta\ -1°$, for instance, is a measure of osmotic value, that $\Delta\ -2°$ is a higher value, and so on.

In real life it would be unusual, indeed impossible, to find such a simple situation, with a solution of a single substance on one side of the membrane and pure water on the other side. There are always solutions of several or many different substances on both sides. Then each solution has its own total osmotic value, Δ, resulting from the combination of all the things dissolved in it. What happens in such a case? If Δ is the same on both sides of the membrane, nothing happens. If it is higher on one side than the other, water will diffuse (predominantly) from the side of *lower* to that of *higher* Δ. In general, the greater the difference the faster the diffusion, or, at any rate, the longer the diffusion will continue before equilibrium is reached.

In this summary of relationships between cells and their environments we have stressed the chemical needs of cells and the varied phenomena of diffusion. It must, however, be realized that any environmental factor may affect cells directly. This may be emphasized by brief consideration of effects of light and heat on cellular processes.

Life requires both materials and energy. Energy takes many forms, one of which is *chemical energy*. Such energy is bound up, stored quietly, so to speak, in chemical compounds. It can be released from such compounds by reactions and turned into other forms of energy, such as heat, movement, or light. Light is produced from chemical energy in an extraordinary array of living things—fireflies and some mushrooms, to mention only two examples. This is, however, a rather minor feature in the vast picture of life as a whole.

Although little of life's energy is turned back into light, practically all of it comes from light in the first place. The few exceptions are of no importance at this point. The process of turning light from the sun into chemical energy, available for all forms of life, is called *photosynthesis* and is a phenomenon of most plants and some protists.

Aside from this, which is by far the most basic effect of light on life, cells are always to some extent affected by light that falls on them. It does not fall on all; most cells of larger animals and plants are internal and in constant darkness. Some whole organisms also live in constant darkness or nearly so: in caves, in the deep sea, in the soil, inside larger plants and animals. Most fungi and bacteria grow best in the dark, and some are killed by much exposure to light.

Cells that do receive light are, at the least, slightly

warmed thereby. Light also causes chemical changes, even in animal cells; this is evident from the apparent effects of light in tan and sunburn. Light also brings about the formation of vitamin D in human and some other animal cells. Bacteria may be killed by light, especially the shorter rays called "ultraviolet," which are the most active in most biochemical processes involving light.

Intensity of light varies greatly, and these variations influence life. In green plants, diffuse light generally favors vegetative growth and, in some plants, bright light favors the formation of reproductive structures. Bright light retards growth in seedlings because of its influence on certain hormones necessary for growth. The daily alternation of light (day) and dark (night) also influences flowering in many plants, migration in animals, and other activities of organisms.

All organisms require some heat in the environment in order to continue functioning. The range of temperatures in which life can exist is usual at the surface of the earth and in its waters. All cells produce some heat incidental to their activities, but as a rule this production is so small that cells are seldom far from the temperature of their environment. All cells also have a range of temperatures, their *optimum* temperature, often quite narrow, in which they do best. The optimum varies considerably in different cells and different organisms, which is a major factor in the distribution of plants and animals in climatic zones.

The reasons (at least the *proximate* reasons, answers to the first "why") for all these relationships lie mainly at the level of the cells. The physical properties of protoplasm, such as its fluidity or elasticity, are considerably affected by its temperature. The rates of biological processes in cells and often also the nature of these processes are also strongly influenced by temperature. Other things being equal, there is a tendency for average rates of these processes to double for every 10° C. rise in temperature. The approximation is quite rough, and there is great variation. There are also processes that reach a maximum rate at a certain temperature and become slower when this point is passed. Thus potato plants synthesize starch more rapidly at 20° than at 30° C., but utilize sugar much more rapidly at the latter temperature. The sugar that is not utilized is stored as starch. Other things being equal, the yield of potatoes is therefore higher at the lower temperature.

It is noteworthy that organic reactions proceed at what seem to the chemist very low temperatures. If you set fire to a lump of sugar in air, it burns with a hot flame. (Doubtless you know the trick of starting this reaction by putting a little cigarette ash on the lump before you touch a match to it.) But sugar is oxidized in us at our body temperature, and at considerably lower temperatures in some other organisms.

Our investigation of the processes within the cell continues with the following article by a great European biologist. As Karl von Frisch indicates, "In the ordinary manner of speaking, when we talk of respiration, we mean breathing in and breathing out. . . . As biologists we mean by respiration not these subsidiary formalities but the essential functions: the oxidative breakdown processes within the cells whereby these obtain their energy for life." A similar statement might be made concerning the functions of the cells in nutrition. Both are here discussed.

A native Austrian, Dr. von Frisch obtained his Ph.D. at the University of Vienna in 1910. He became Professor of Zoology at the University of Rostick in 1921, and also taught at Breslau and Graz, Austria. His book The Language of Bees *is a charming volume and is considered the authoritative work on the subject. The recipient of many honorary awards, he has been Professor Emeritus of the Zoological Institute of the University of Munich since 1950. "Nutrition and Metabolism" is a selection from Dr. von Frisch's* Life, *a biological treatise which has recently been published in an English translation.*

NUTRITION AND METABOLISM

KARL VON FRISCH

A LARGE PART of food is continually used as fuel. In this aspect of their metabolism, men and animals resemble machines that can accomplish something only if they are heated or given energy in some other way. Yet in living creatures we seek in vain for heating apparatuses and hissing flames. The chemical processes that enable us to work and to maintain our body heat are performed inconspicuously in the internal parts of all living cells. Nevertheless, they are processes essentially comparable to combustion. If an old-fashioned steam engine pulls a train, the energy for its work is derived from the fuel that is burned, from wood, for instance, if it is old enough. Its carbohydrate (cellulose) is transformed, when oxygen is added (oxidation), to the simple compounds CO_2 and H_2O, and energy becomes free, appearing as work and as heat. When wood is burned, the development of heat is so violent that the particles (gases) evaporated by the heat glow and form flames. When a horse pulls a wagon, chemical transformations provide the energy. In these, large molecular compounds, especially glycogen in the muscle cells, are broken down to simpler compounds with the uptake of oxygen, and finally to CO_2 and H_2O. The combustion process in the living cells is much more complicated, but as in the wood fire, energy is released and appears as work and heat—in the horse as well as in a man or in a bee.

Animal and Plant Metabolism: Consumers and Producers, the Great Opposites

Oxygen is necessary for the completion of the oxidation processes just described. In this way also the body behaves like a heated machine. If the supply of air is cut off in a steam boiler, the fire is stifled, and after a short time the machine stops. If the source of air for our bodies is cut off, we too suffocate. By breathing air we continually provide ourselves with the amount of oxygen we need.

In the ordinary manner of speaking, when we talk of respiration, we mean breathing in and breathing out. When we breathe in, fresh air is brought into the lungs and from there the oxygen is transported by the blood to the whole body. When we breathe out, the air that has been used and loaded with CO_2 is eliminated. As biologists we mean by respiration not these subsidiary formalities but the essential functions: the oxidative breakdown processes within the cells whereby these obtain their energy for life. This holds true not only for man but also very generally for the cells of animal and plant bodies. They all respire, and in the process molecular compounds are split when oxygen is taken up, until finally CO_2 and H_2O are produced.

Where does this finally lead? If plants and animals obtain their vital energy in such a way that they are continually splitting up and destroying their fuels and reserves, then the process must at some time come to an end; every supply depot must some day become depleted if things are only removed and nothing put in.

In fact, after a short time all life on earth would die away if some higher power were to annihilate the green plants. They are the producers of organic substances, the opposites of the hungry animal world. They construct what the animals destroy and they keep life in equilibrium. They alone are able to capture enough of the great supply of energy that the sun beams down, day in and day out, on our earth; they alone can apply it to the construction of complicated compounds on which not only they themselves, but all animals and plants can live.

The green plants owe their color to chlorophyll.[1] This pigment, closely related chemically to the red pigment of blood, is bound to microscopically small granules in the plant cells, the chloroplasts.[2] These are present only in illuminated parts

1. *Chloros* (Greek) = "light green"; *phyllon* (Greek) = "leaf."
2. *Plastis* (Greek) = "modeler," "shaper."

of plants. They are absent from the roots in the dark soil. In the absence of light, chlorophyll does not develop in parts above the earth. Asparagus raised under flowerpots remains as white as does the seed of a plant germinating in a dark cellar. Even these simple experiments show that chlorophyll has something to do with light. Pigments appear colored to us because they absorb a particular portion of the spectrum of light rays. Chlorophyll appears green because it absorbs mostly red and blue rays, but it allows green light to penetrate freely without absorbing it. The energy of the captured light is utilized and serves in the construction of highly complicated compounds from simple building materials. During this process energy is stored and oxygen is released (reduction processes). The energy can be used again later when the compounds are oxidized (respiration). The chemical details of the construction of the organic compounds are still to a high degree a factory secret of the plant cells. Here, in rough outline only, we shall consider what happens.

We know that the higher plants take nutrients out of the soil with their roots. It is difficult to determine what, under natural conditions, they actually take out. If a plant is cultivated, however, in an artificial solution of nutrients, in a vessel of water to which known and weighed amounts of certain chemicals have been added, then it can easily be determined what the plants require for life. They can grow excellently and can flourish if certain inorganic substances are present in the nutrient solution: saltpeter (KNO_3), calcium sulfate ($CaSo_4$), magnesium sulfate ($MgSO_4$), phosphate [$Ca_3(PO_4)_2$ and $Fe_3(PO_4)_2$], and traces of still a few other elements. In this experiment we see growing before our very eyes a plant body that is constructed essentially of organic compounds, thus of carbon compounds (carbohydrates, fats, and proteins). Since no carbon has been provided in the nutrient solution, the plant can only have taken this out of the air. This contains carbon dioxide (CO_2) in very small but constant amounts (about $^1/_{1000}$th of a gram in two liters of air). Air enters through the stomata and gets inside the leaf. Here the plant takes the carbon dioxide from the air and uses its carbon to form its own carbon compounds. This synthesis of organic compounds is called assimilation. Energy is used in the process and it is stored in the organic compounds that are formed. The required energy is taken by the plant from the sunlight when the chlorophyll absorbs part of the light.

It used to be believed that during assimilation carbon dioxide is broken down into its components C and O_2, the carbon used, and the oxygen given off to the outside. It has, however, been proved that the process is much more complicated. The CO_2 is first bound to an organic molecule. Only in this bound form is it reduced, by hydrogen which has been removed from water molecules. This takes place in the chloroplasts, under the influence of light and with the assistance of enzymes. From the OH radicals that remain after the removal of the hydrogen, there develops during assimilation free oxygen that thus is derived from water and not from CO_2:

$$4\,(OH) \rightarrow 2\,H_2O_2 \rightarrow 2\,H_2O + O_2.$$

With the mediation of specific enzymes, the hydrogen reduces the bound CO_2, leading to the formation of glucose ($C_2H_{12}O_6$) through a series of intermediate products over chemical pathways that are only partially known. Phosphoric acid plays an important role in these processes. When the glucose molecules become combined with each other through the loss of water, the polysaccharide starch $(C_6H_{10}O_5)_n$ is formed. This can be seen in the chloroplasts at the first microscopically visible product of assimilation. The organic substances that have been produced are next stored in the leaf and thence conducted further to other parts of the plants according to need; they are again broken down to soluble glucose for transport.

A hundred-year-old tree contains about one and a quarter tons of pure carbon, which during its lifetime it has captured and concentrated from the scanty amount of carbon dioxide in the air; in order to accomplish this, gigantic amounts of air must have been worked over. It is easy to see why a tree has so many leaves.

Thus in the light the green plants assimilate, taking in CO_2 from the air and giving O_2 to it. But in addition, the opposite processes are also continually going on in the plants, because plant cells, exactly like animal cells, must obtain from chemical transformations the energy to carry on their vital processes. As in animals, this occurs through respiration and the oxidative breakdown of organic compounds to CO_2 and H_2O. Therefore the green plants also respire, using O_2 and expelling CO_2. In daylight the assimilation greatly predominates over respiration and masks it. Not only are the organic compounds constructed that in time will be oxidized; but—to express it in terms of storage economy—extra amounts are prepared, just as on a well-to-do farm more food is raised than the particular farmer's household needs. But at night, in the dark, assimilation ceases, and then respiration, the use of oxygen and the release of carbon dioxide, is easy to demonstrate even in green plants.

So far only the synthesis of carbohydrates has been discussed. These play a significant role in the metabolism of plants. Sugar is the principal fuel of plants, and starch their most important storage material, while cellulose builds their cell walls and their whole supporting structure. Just as in animals, so also in plants, fats can be made from carbohydrates; here the accomplishments of plants do not excel those of the animal body.

On the other hand plants prove themselves successful manufacturers when they form proteins, and for this even the animal world is dependent upon them for its existence. The higher plants take from the soil with their roots, in the form of minerals, the nitrogen necessary for the formation of proteins; we must always add mineral nitrogen to the nutrient solution used in water culture (hydroponics) if our plants are to thrive. In sandy soil that lacks nitrogen compounds, growth comes to a standstill as soon as the materials stored in the seed are used up. Plants cannot utilize the free nitrogen that is so richly at their disposal in the air. From the simple nitrogen compounds taken in by the roots, and from the carbohydrates formed by the green parts, the plants create proteins, which no animal can make for itself from such elementary constituents.

The whole animal kingdom subsists on the organic compounds formed by green plants. This can be seen directly if we observe cows at pasture, or a caterpillar feeding on a leaf. When a fox seizes a hare, it is consuming animal nutriment, but this is only a short detour, because the hare has grown up eating cabbage leaves and other greens. Sometimes the paths are more roundabout; the nutriment can go through a whole series of animals' stomachs, but whenever we follow the chain back to the beginning we come to the plants, the true producers.

There is a continuous cycle of materials in the world of life. The organic compounds built by plants serve not only as nutriment for the plants themselves but also for animals. The animals live on the substances plants have made and destroy them; in breaking down these compounds, they also use the oxygen that assimilating plants have given off. The end products of animal metabolism, carbon dioxide, and simple nitrogen compounds that are returned to the soil by animals in their urine and feces, are in their turn nutriment for plants. Whatever organic compounds are stored in plant and animal bodies during their lives are after death restored to the great cycle when the bodies die and are broken down and destroyed by the bacteria of putrefaction.

These bacteria belong to the lower plant world, and yet they usually participate in organic life by destroying rather than by building. It would be a mistake to think of all plants as opposites of the animals. Not only bacteria but also molds and fungi, and some higher plants that lead a parasitic existence, lack chlorophyll, and are, like animals, consumers, not producers, of organic material. Only the green plants are the manufacturers of organic compounds. As autotrophic[3] organisms, they maintain the balance of life for the heterotrophs.[4]

Since the green plants do not generally move about, they can flourish only if they can find close at hand all the substances necessary to construct their bodies. They need have no worry about a sufficient supply of carbon. Carbon dioxide, a highly mobile gas, is distributed so evenly throughout the air that in spite of its small proportion the amounts necessary for life are everywhere available. It is different with the

3. *Autos* (Greek) = "self"; *trephein* (Greek) = "to nourish."
4. *Heteros* (Greek) = "different," "other."

mineral substances that are taken out of the soil. Yet even here the economy remains in order so long as man does not unbalance things. Under natural conditions when plants die and when animal corpses deteriorate, all the mineral components incorporated in them are given back again to the soil, but in meadows and plowed lands used in agriculture this does not happen. The nutrients that the growing grass has removed from the soil and that man has made use of after a detour through cattle via milk or meat, these are removed forever from the meadowlands. Similarly the regular removal of grains at harvest time can rapidly deplete the land of nitrogen, phosphorus, potassium, and other mineral substances essential for plant growth. The farmer must artificially re-establish the equilibrium. He can replenish the deficient soil at least in part by using solid or liquid dung or compost. These natural organic fertilizers offer the advantage that they also promote the formation of humus and of loose crumbly soil. But they do not return to the earth anywhere nearly everything that has been removed from it. Today, therefore, when land is intensively cultivated, an appreciable quantity of mineral fertilizers (artificial fertilizers) must be added. Above all, nitrogen, phosphorus, and potassium must always be furnished to the soil in order to maintain high harvest yields.

An aquarium keeper regards with proud pleasure the balanced economy of his aquarium if this is correctly stocked and well planted. Here, in a small space, the same cycle is in progress which occurs in increased measure in a lake, in an ocean, or in the atmosphere. The narrow walls of the aquarium provide no space for the reserves that exist in nature, and any incorrect proportion between plants and animals becomes perceptible quickly and inexorably. If the beginner, in his enjoyment of active creatures, puts in too many wriggling and swimming animals, then the fish snap for air at the surface because they need oxygen, and the water becomes foul and cloudy because the excretions of the animals exceed by far the nutritive needs of the plants. But if the proportions are correct, equilibrium prevails, and in a narrow space animals as well as plants live, so to speak, from hand to mouth, and each in its turn uses up what the other gives away.

The exchange proceeds actively in nature too. It has been calculated that all the carbon dioxide in the earth's atmosphere would be completely used up in a few years by the assimilation of plants if it were not for the fact that it is continually formed anew and given back to the atmosphere by respiring organisms. It is a hungry company of animals and plants that populates our earth, and anything usable does not have to wait long to be taken advantage of. Sometimes, however, it does not happen so fast. The wood-beetle grub that lives in the beams of an old farmhouse is benefiting from energy supplies that were stored up perhaps centuries before when those beams were still trees standing in the sun. When we warm ourselves before a coal fire, and so help ourselves to maintain our own body heat, we are using energy that our ancient sun radiated down on the plant world of the Carboniferous era more than three hundred million years ago and that was bound by these plants.

In previous selections we have discussed the history of the discovery of cells and some of their functions, and we have seen how all organisms, from the largest to the smallest, engage in certain activities in common. In the following article R. W. Gerard describes these activities as they occur in an animal consisting of only a single cell. The amoeba reacts to its surroundings; it discovers, ingests, and digests food; it somehow senses the need for actions essential for its preservation; the scene of constant and intricate chemical reactions, it nevertheless retains its individuality; it slowly grows, and on reaching a critical size, it divides and reproduces. In so doing, it performs functions similar to those of organisms billions of times its size. The life span of the amoeba, for all its relative simplicity, follows a universal pattern.

Dr. Gerard became a Ph.D. at the University of Chicago in 1921 and an M.D. at Rush Medical College in 1924. He has been an editor of a number of physiological journals and has done research at institutions in both the United States and Europe. He is now Professor of Neurophysiology, Mental Health Research Institute, the University of Michigan. "Odyssey of an Amoeba" is a chapter from his book Unresting Cells.

ODYSSEY OF AN AMOEBA

R. W. GERARD

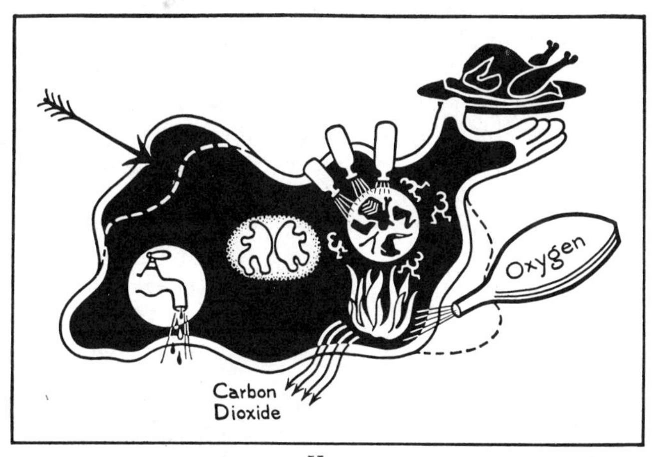

II-1.

FOCUS a microscope on a drop of blood obtained from a finger prick. The color disappears and the opaque fluid becomes a crowd of disclike bodies, pale orange, piled in irregular masses or floating singly in a colorless liquid. Smooth, round, flat, these "red" corpuscles of the blood are powerless barges jostled about by the currents in the plasma sea. Present in enormous numbers, vital to our life, they do not now concern us except as they obscure their paler fellows. The white cells (leucocytes) will be overlooked without search, for they are rare, a scant seven thousand in a cubic millimeter or one hundred million odd in a cubic inch, compared with seven hundred times as many reds. They will be found more easily in pus, which is made largely of their bodies. But here comes one into view; not floating freely, but clinging to the surface of the glass plate on which the drop is placed. The leucocyte's bulging outline is sprawled out, with several protuberances stretching away from the main mass. Within this margin is a clutter of many small grains that move irregularly and one larger lump that seems at rest. The whole is not half of one-thousandth of an inch across, yet its parts are clearly visible.

Not much appears to go unless the white cell is kept warm, but a light bulb placed close to the microscope supplies adequate heat. Now the leucocyte "comes to life." One of the bulges (really "false foot" or "pseudopod") stretches out, the content of the cell flows into it, the main mass squeezes in behind and soon the peninsula contains the whole mass. Another bulge appears, again the protoplasm crowds into it. The leucocyte is slowly crawling on the glass. Soon it is lost to view beneath the red corpuscles.

It is not difficult to obtain the white cells separated from the others, and immersed in plasma, the fluid part of blood. Let us take such a drop of plasma with leucocytes floating in it, add some typhoid bacilli scraped from a mass growing on jellied agar, mix them well, and keep the whole warm. After

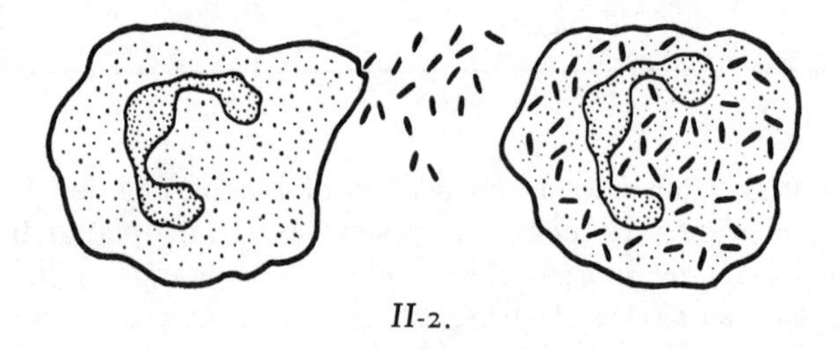

II-2.

some time the bacilli are seen, of course through the microscope, to be within the cells. It matters little if the typhoid germs be alive or previously cooked to death, the leucocyte devours them with equal ease. Often, a hundred of the tiny sausages make a comfortable meal for one white cell. True, if the germs are alive to start with they remain alive for some time after ingestion, and it may fare ill with the white whale carrying an overload of Jonahs. Within the body, the leucocyte's life is of little import if it but discharge its duties of defending against bacteria and helping to clear up all sorts of tiny particles of debris. But we are interested in the leucocyte as a tiny living animal, one, in fact, closely related, in much that it does and is, to the familiar amoeba of pond water. Bacteria or any other tiny bits of living or formerly living matter are just so many good morsels to be eaten, digested, and used for the needs of the cell.

We can spy on the leucocyte, or more easily the amoeba, obtaining its meal. Here is a small amoeba crawling slowly along the glass slide by means of its pseudopods. In one direction lies a tiny green filamentous plant, a string of protoplasmic masses enclosed in "wooden" boxes. The amoeba oozes here and there, without seeming to make a course for anywhere in particular; yet fairly soon it somehow manages to reach the thread. Is this pure luck; or the principle that by moving about at random for sufficient time one must cover the available territory and run into whatever is in it; or is the amoeba somehow guided to the food? With patience, we could settle this by watching one amoeba after another and carefully noting the course each one takes,

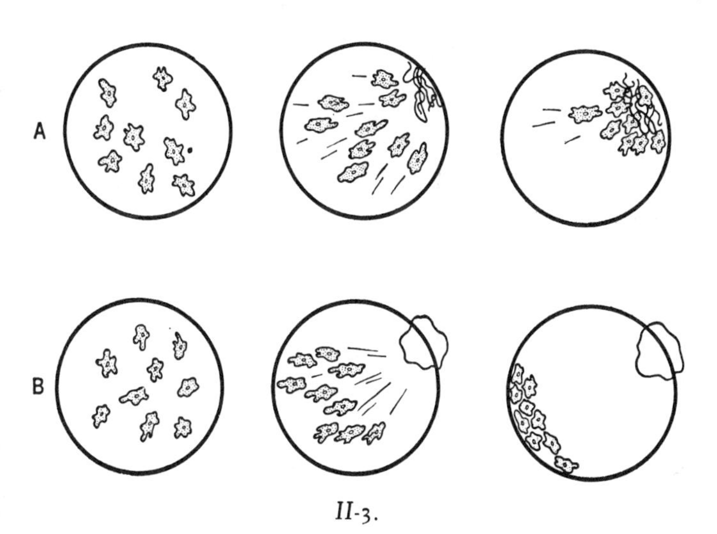

II-3.

but it is much simpler to place a large number of amoebae at random in a large drop and then introduce some food material near one edge. We need not watch long. Almost at once a population drift toward the food is present and soon the whole amoeba mob is clustered densely around the plants. Clearly they have somehow been directed, and since there is every reason to doubt their possession of any means of seeing, smelling, or otherwise becoming "aware" of the object, the cue must be so simple and direct that it is almost compelling.

The little beasts are "irritable" in the sense that environmental differences make them move toward one condition or from another. In this case, minute amounts of chemical substances which are produced by the plants and spread through the water serve to direct the amoebae. If, instead of the plants themselves, a bit of juice squeezed from them, or even water in which they have been kept, is placed somewhere in the amoebae's pool, these hungry mites will still gather in this region. Other substances, such as weak vinegar, will similarly drive the animalcules away from them; which is a very useful response, since vinegar would soon kill the cells if they remained. But we dare not conclude from this that the amoebae are being intelligent, for a moderate electric current, which is just as fatal as the vinegar, draws the amoebae along it before exploding them.

No, all these substances force the movements of the animal, in part, by making that part of its surface which is nearest to them more, or less, tense than the opposite one. If the surface facing the vinegar becomes more tense than the far one—if, that is, its surface tension increases—then the bulk of the fluid inside is forced from this side to bulge out the other and the whole drop will move away. The reverse occurs in the presence of substances from food, for these make the near surface give way, lower its surface tension, and the amoeba advances to its meal. In the same way, a drop of alcohol on a sheet of glass moves toward a warmed region while a drop of oil moves from it. A simple mechanism, this, for getting the amoeba into desirable and out of dangerous places (for an amoeba would ordinarily

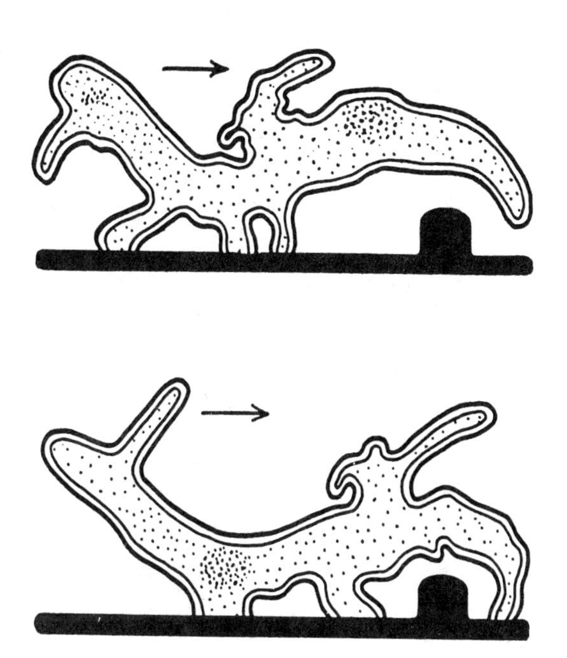

II-4.

never come into the presence of considerable currents, though it might easily encounter the vinegar); yet perhaps not so different from the equally automatic devices that make a man sneeze away an irritant that tickles a hair in his nostril. And, in truth, not so simple either, as we discover on studying our animal with even finer tools than the microscope.

We return to our peephole just in time to see the hungry amoeba, which has by now sprawled itself over to its prospective meal, give an amazing performance of sword-swallowing. There is no mouth, of course, or any ready-made stomach into which to swallow the plant; for that matter, the thread may be many times as long as its captor. No matter. The protoplasmic blob reaches and flows around an end of the woody filament and soon surrounds it like a lollypop on a stick. Then, as the amoeba crawls still farther, the thread bends, coils on itself, and finally becomes wound up tightly within its living sarcophagus. Nothing more seems to happen, and we might imagine the little beast sleeping peacefully after a large dinner. Really the work is just beginning, for the amoeba has only ingested its meal and has yet to digest it.

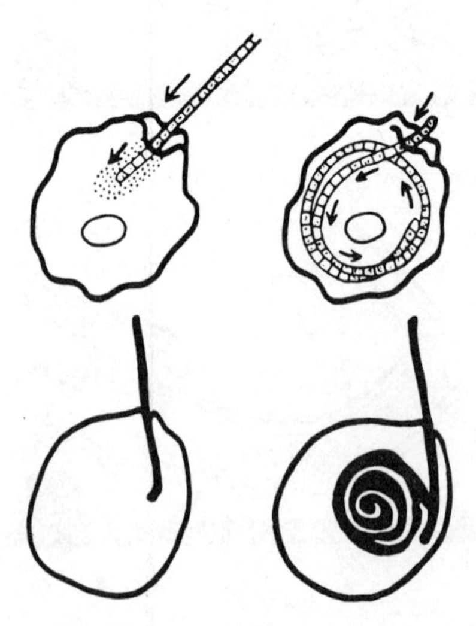

II-5. *Amoeba eating filament of algae (above), and a chloroform drop "eating" a thread of shellac. The visible events look very similar. (Partly after Rhumbler.)*

While the amoeba now remains still, we can turn the high power of the microscope upon its fairly transparent body and watch this process of digestion. Possibly, when the alga thread was engulfed, a bit of water was included with it, but whether this was so or not, the coiled strand is soon seen to be clearly separated from the amoeba's protoplasm. It lies suspended in a tiny, perfectly clear droplet of fluid, a vacuole, the walls of which are just the protoplasm of the amoeba, even though they were originally from the slightly thickened protoplasmic surface in contact with outside fluids. The vacuole develops around the food mass wherever that may happen to lodge and not at some prearranged portion of the tiny anatomy. This simple arrangement is all the stomach the amoeba has with which to work.

We continue to watch but see no further movements; the walls of the vacuole do not churn or grind and so mechanically break up the food, as do the muscular walls of more complicated digestive systems; nothing at all seems to be happening. Yet, slowly, changes are occurring, even visibly, for the sharply outlined green algal spiral becomes blurred, the green disappears, the fine granules and rods present within its protoplasm smudge and vanish, and eventually nothing is left of the plant cells but their thin, tough walls of wood. This undigestible and useless carcass is finally eliminated from the amoeba by the same straightforward technique which led originally to its ingestion. The amoeba's protoplasm flows forward until the residue comes to lie against the hindmost surface, which opens and returns it

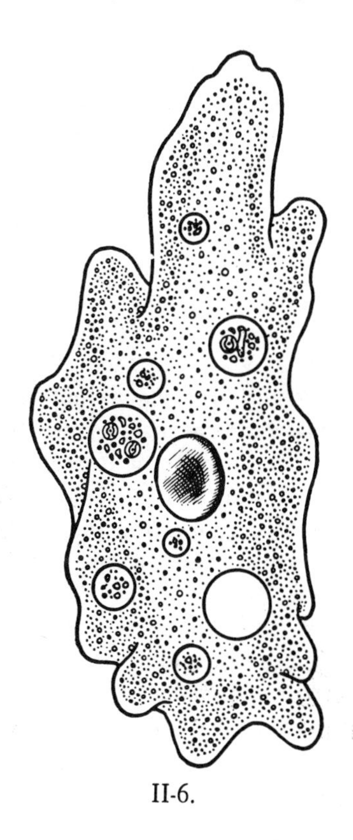

II-6.

to the great outside, while the amoeba sluggishly goes on its way.

Somehow, while it remained in the vacuole, food was digested and the useful substances resulting from this were absorbed or taken into the protoplasm of the amoeba. Yet we saw no movement during this entire time. Clearly the vacuole fluid was vastly different from the original pond water in which the algae, far from disintegrating, grew and flourished. Into the vacuole stomach must have come from the amoeba's protoplasm, along with the water, some dissolved substances able to act upon and digest the protoplasmic materials of the algae. Such digestive juices are present in all stomachs, and when large quantities are available, as, for example, in our own, it is no great problem to collect enough juice for study in the test tube. And, right enough, if a piece of meat or a bit of hard-boiled egg is placed in this gastric juice, it is quickly seen that the food softens, breaks up, and finally disappears, while the originally clear liquid becomes for a time cloudy and opalescent before it again turns perfectly transparent. During the same time, like bits of food placed in an equal amount of water remain solidly unchanged. But now we have satisfied ourselves that somehow the complex insoluble substances present in the food have been torn apart, chemically dismembered, and reduced to smaller, simpler, soluble substances, which, like sugar dissolved in water, are invisible, and which, by passing easily along with the water molecules, can diffuse out from the vacuole into the mass of protoplasm, which is the amoeba's body. This is the true importance of the whole complex process of finding, eating, and digesting a meal; to bring these dissolved molecules, the fragments of food, into the protoplasm of the cell. It is here that they become useful to the animal. And it seems likely enough that the substances produced by breakdown of the alga's protoplasm could be effectively used in building up that of the amoeba. But even now, when the amoeba has not merely digested its food but even disposed of the detritus, the work is really only well begun.

In the body protoplasm, of the single cell which constitutes an amoeba or of the myriad cells which make up a man, a cold flame is steadily burning. The chemical turmoil and traffic called metabolism somehow leaves the main characteristics of the protoplasm in which it occurs essentially constant. Of course, stimuli produce temporary changes, but the system always reverts toward some steady state, much as a candle flame flickers in a passing draft but regains its even glow when the air subsides. In the candle flame the metabolism is destructive; the candle fats are burned by oxygen to carbonic acid, and the candle slowly disappears. In protoplasm, likewise, fats and sugars and albumin are also being constantly burned away. Strangely though, the animate candle here does not shrink, but grows. Not only are the combustible elements of food and protoplasm being continuously destroyed by disruptive metabolism, catabolism, but more rapidly than this the protoplasm is being rebuilt by constructive metabolism, anabolism, from a continuous new supply of food. Stop the food and the amoeba, like man, will starve to death, while slowly burning up its own body to keep going.

This is reminiscent of the famous race, almost a half-century ago, of the Hudson River steamers, the Oregon and the C. Vanderbilt. The Oregon, with fires blazing and paddles churning, entered the last lap of the race well ahead of its rival but with no more fuel. The wooden furniture, berths, all dispensable parts of its structure, were torn up and fed to the furnace until the emaciated hulk paddled victorious into the harbor off the Battery in New York City. So the starving amoeba, whose protoplasm contains much the same material that is present in its food, burns its decks and finally its very hull, and is gone. But again the comparison with man-made machines, as with candles, breaks down, for the protoplasm continually rebuilds itself. Repair not only keeps pace with destruction but, during much of the time in most living things, building is more rapid than wear, and growth results.

Most of this subtle alchemy is quite invisible and must be unraveled indirectly; but its consequences can be seen just as clearly under the microscope as was the earlier hegira of the hungry amoeba in search of a dinner. We have watched this animalcule, only a hundred-fiftieth of an inch long, crawl to food, engulf particles, digest them, extrude their solid residue, and go its way seemingly quite unchanged. But if we measure the cell from time to time, a slow growth in size is readily detected. When it has almost doubled its volume (not its length) there occurs a series of bizarre events indeed. To do them justice, they must really be followed under the microscope or, even better, seen in speeded-up motion pictures taken through a microscope; for words convey but little of the glory of a sunrise, and here is one in miniature. The delicate, intricate, and entirely regular maneuvers through which all elements of the cell pass in changing from one large parent amoeba to two small daughter ones constitute one of the grand strategies of life. Practically all living cells that grow and reproduce follow this general pattern of cell division, mitosis (forming threads), with unbelievable faithfulness.

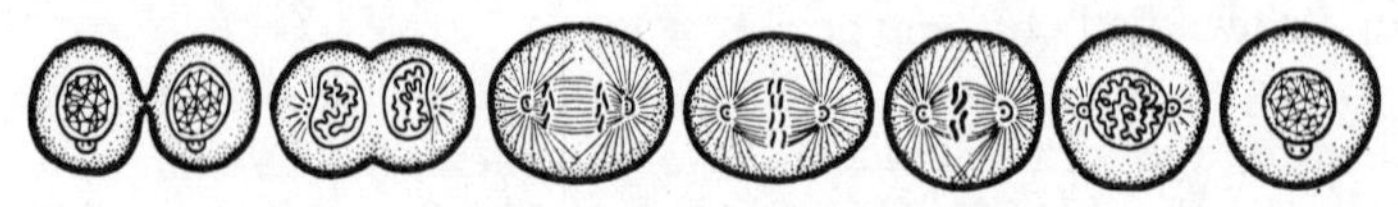

II-7. *From above to below, successive stages in a typical cell division by mitosis. The really crucial step is from the third to the fourth stage, when each of the chromosomes lined up on the spindle divides into two equal halves.*

First the nucleus, which was a clear, practically structureless fluid droplet, separated from the surrounding cell protoplasm by a distinct membrane, begins to cloud up. The cloud condenses slowly and takes the form of a much-entwined thread of more solid material called chromatin, while the membrane around becomes blurred and broken and finally disappears. While these changes within the nucleus are progressing, a tiny granule at one side, tiny even compared to the nucleus, divides and the two halves move so as to occupy opposite poles of a diameter through the nucleus. Beginning as they move and continuing after they are in their new positions, there grow mysteriously from them as centers two rosettes of gossamer threads that radiate in all directions to the farthest limits of the cell. At this stage the picture is like that of a Christmas card showing beams of light radiating from a star. By the time these two "asters" (stars, as they are appropriately called) have fully formed, the nuclear membrane is entirely gone (amoeba is rather nonconformist here) and the chromatin thread has shortened and thickened and divided into some determined number of stubby bars (chromosomes) which line up in the equator between the asters. The rays pointed toward them become more prominent and gather together so that a clear spindle is seen.

One might say that up until now there is still one cell; the next step is definitely over the line, for now each chromosome separates along its length into two exactly equal halves. "Exactly" is true to an unbelievable degree, for the daughter cells in turn divide, their offspring do likewise, and so on; yet, the chromosome halves that are formed during each division remain, generation after generation, exactly equal. One-half of each chromosome now rapidly moves toward its own end of the spindle, leaving the center more or less empty. Then the whole process unwinds itself; the rays of the asters shrink and disappear, each group of chromosomes stretches again into a contorted thread, which becomes longer and thinner until it breaks again into a cloudy mass, and a new membrane forms around it. Meanwhile the cell surface becomes indented along the circumference midway between these two nuclei, and the cleft becomes ever deeper until the whole mass of protoplasm has been pinched in two. The two daughter amoebae separate and wend their individual ways in search of another meal.

Somehow, out of the foods, protoplasm has built itself. Somehow, all the individual structures, substances, and parts of the cell have increased and multiplied. Somehow, one individual has become two and like has produced like. How these marvelous chemical and architectural feats are achieved constitutes a basic problem of the science of life and one to which there are but partial answers. We may be confident, however, that so long as we keep the amoebae in enough fluid with food and oxygen available, generation after generation of them will live out their simple lives—laboriously crawling away from danger and toward food, ingesting and digesting their tiny meals, stoking the chemical flame of life, excreting the waste, growing to maturity, and finally reproducing in kind a new generation to repeat the cycle.

III. The Origin of Life

III. *The Origin of Life*

Next to organic evolution, a subject to be discussed in a later section, the most hotly debated biological question has been that of the nature and origin of life. Unlike evolution, which first received wide attention in 1859, vitalist theories have existed throughout history. Scientists have approached the problem gingerly, as being outside their province; but occasional experimenters have attempted explanation in physical and chemical terms. Galvani, for example, believed that the current of electricity which caused a frog's leg to twitch convulsively was the "vital force" for which mankind had been searching. As knowledge advanced, he was proved in error; indeed the whole history of vitalism has been one of retreat before scientific progress. The controversy over spontaneous generation has been one of the most important chapters of this story. Its climax occurred with a series of experiments in which Louis Pasteur conclusively destroyed the theories of the spontaneous generationists. Pasteur was, of course, one of the greatest scientists of the nineteenth century. He was the subject of a classic biography written by his son-in-law, from which the following selection is taken. It describes the background and nature of the experiments with gusto.

LOUIS PASTEUR AND SPONTANEOUS GENERATION

RENÉ VALLERY-RADOT

ON JANUARY 30, 1860 the Académie des Sciences conferred on Pasteur the Prize for Experimental Physiology. Claude Bernard, who drew up the report, recalled how much Pasteur's experiments in alcoholic fermentation, lactic fermentation, the fermentation of tartaric acid, had been appreciated by the Académie. He dwelt upon the great physiological interest of the results obtained. "It is," he concluded, "by reason of that physiological tendency in Pasteur's researches, that the Commission has unanimously selected him for the 1859 Prize for Experimental Physiology."

That same January, Pasteur wrote to Chappuis: "I am pursuing as best I can these studies on fermentation which are of great interest, connected as they are with the impenetrable mystery of Life and Death. I am hoping to mark a decisive step very soon by solving, without the least confusion, the celebrated question of spontaneous generation. Already I could speak, but I want to push my experiments yet further. There is so much obscurity, together with so much passion, on both sides, that I shall require the accuracy of an arithmetical problem to convince my opponents by my conclusions. I intend to attain even that."

When Biot heard that Pasteur wished to tackle this study of spontaneous generation, he interposed, as he had done seven years before, to arrest him on the verge of his audacious experiments on the part played by dissymmetrical forces in the development of life. Vainly Pasteur, grieved at Biot's disapprobation, explained that this question, in the course of such researches, had become an imperious necessity; Biot would not be convinced. But Pasteur, in spite of his quasifilial attachment to Biot, could not stop where he was; he had to go through to the end.

"You will never find your way out," cried Biot.

"I shall try," said Pasteur modestly.

Angry and anxious, Biot wished Pasteur to promise that he would relinquish these apparently hopeless researches. J. B. Dumas, to whom Pasteur related the more than discouraging remonstrances of Biot, entrenched himself behind this cautious phrase—

"I would advise no one to dwell too long on such a subject."

Senarmont alone, full of confidence in the ingenious curiosity of the man who could read nature by dint of patience, said that Pasteur should be allowed his own way.

It is regrettable that Biot—whose passion for reading was so indefatigable that he complained of not finding enough books in the library at the Institute—should not have thought of writing the history of this question of spontaneous generation. He could have gone back to Aristotle, quoted Lucretius, Virgil, Ovid, Pliny. Philosophers, poets, naturalists, all believed in spontaneous generation. Time went on, and it was still believed in. In the sixteenth century, Van Helmont—who should not be judged by that one instance—gave a celebrated recipe to create mice: any one could work that prodigy by putting some dirty linen in a receptacle, together with a few grains of wheat or a piece of cheese. Some time later an Italian, Buonanni, announced a fact no less fantastic: certain timberwood, he said, after rotting in the sea, produced worms which engendered butterflies, and those butterflies became birds.

Another Italian, less credulous, a poet and a physician, Francesco Redi, belonging to a learned society calling itself The Academy of Experience, resolved to carefully study one of those supposed phenomena of spontaneous generation. In order to demonstrate that the worms found in rotten meat did not appear spontaneously, he placed a piece of gauze over the meat. Flies, attracted by the odour, deposited their eggs on the gauze. From those eggs were hatched the worms, which had until then been supposed to begin life spontaneously in the flesh itself. This simple experiment marked some progress. Later on another Italian, a medical professor of Padua, Vallisneri, recognized that the grub in a fruit is also hatched from an egg deposited by an insect before the development of the fruit.

The theory of spontaneous generation, still losing ground, appeared to be vanquished when the invention of the microscope at the end of the seventeenth century brought fresh arguments to its assistance. Whence came those thousands of creatures, only distinguishable on the slide of the microscope, those infinitely small beings which appeared in rain water as in any infusion of organic matter when exposed to the air? How could they be explained otherwise than through spontaneous generation, those bodies capable of producing 1,000,000 descendants in less than forty-eight hours.

The world of salons and of minor courts was pleased to have an opinion on this question. The Cardinal of Polignac, a diplomat and a man of letters, wrote in his leisure moments a long Latin poem entitled the *Anti-Lucretius.* After scouting Lucretius and other philosophers of the same school, the cardinal traced back to one Supreme Foresight the mechanism and organization of the entire world. By ingenious developments and circumlocutions, worthy of the Abbé Delille, the cardinal, while vaunting the wonders of the microscope, which he called "eye of our eye," saw in it only another prodigy offered us by Almighty Wisdom. Of all those accumulated and verified arguments, this simple notion stood out: "The earth, which contains numberless germs, has not produced them. Everything in this world has its germ or seed."

Diderot, who disseminated so many ideas (since borrowed by many people and used as if originated by them), wrote in some tumultuous pages on nature: "Does living matter combine with living matter? how? and with what result? And what about dead matter?"

About the middle of the eighteenth century the problem was again raised on scientific ground. Two priests, one an Englishman, Needham, and the other an Italian, Spallanzani, entered the lists. Needham, a great partisan of spontaneous generation, studied with Buffon some microscopic animalculae. Buffon afterwards built up a whole system which became fashionable at that time. The force which

Needham found in matter, a force which he called productive or vegetative, and which he regarded as charged with the formation of the organic world, Buffon explained by saying that there are certain primitive and incorruptible parts common to animals and to vegetables. These organic molecules cast themselves into the molds or shapes which constituted different beings. When one of those molds was destroyed by death, the organic molecules became free; ever active, they worked the putrefied matter, appropriating to themselves some raw particles and forming, said Buffon, "by their reunion, a multitude of little organized bodies, of which some, like earthworms, and fungi, seem to be fair-sized animals or vegetables, but of which others, in almost infinite numbers, can only be seen through the microscope."

All those bodies, according to him, existed only through spontaneous generation. Spontaneous generation takes place continually and universally after death and sometimes during life. Such was in his view the origin of intestinal worms. And, carrying his investigations further, he added, "The eels in flour paste, those of vinegar, all those so-called microscopic animals, are but different shapes taken spontaneously, according to circumstances, by that ever active matter which only tends to organization."

The Abbé Spallanzani, armed with a microscope, studied these infinitesimal beings. He tried to distinguish them and their mode of life. Needham had affirmed that by enclosing putrescible matter in vases and by placing those vases on warm ashes, he produced animalculae. Spallanzani suspected: firstly that Needham had not exposed the vases to a sufficient degree of heat to kill the seeds which were inside; and secondly, that seeds could easily have entered those vases and given birth to animalculae, for Needham had only closed his vases with cork stoppers, which are very porous.

"I repeated that experiment with more accuracy," wrote Spallanzani. "I used hermetically sealed vases. I kept them for an hour in boiling water, and after having opened them and examined their contents within a reasonable time I found not the slightest trace of animalculae, though I had examined with the microscope the infusions from nineteen different vases."

Thus dropped to the ground, in Spallanzani's eyes, Needham's singular theory, this famous vegetative force, this occult virtue. Yet Needham did not own himself beaten. He retorted that Spallanzani had much weakened, perhaps destroyed, the vegetative force of the infused substances by leaving his vases in boiling water during an hour. He advised him to try with less heat. But both on Needham's side and on Spallanzani's there was a complete lack of conclusive proofs.

On December 20, 1858, a correspondent of the Institute, M. Pouchet, director of the Natural History Museum of Rouen, sent to the Académie des Sciences a *Note on Vegetable and Animal Proto-organisms spontaneously Generated in Artificial Air and in Oxygen Gas.* Pouchet, declaring that he had taken excessive precautions to preserve his experiments from any cause of error, proclaimed that he was prepared to demonstrate that "animals and plants could be generated in a medium absolutely free from atmospheric air, and in which, therefore, no germ of organic bodies could have been brought by air."

On one copy of that communication, the opening of a four-year scientific campaign, Pasteur had underlined the passages which he intended to submit to rigorous experimentation. The scientific world was discussing the matter; Pasteur set himself to work.

A new installation, albeit a summary one, allowed him to attempt some delicate experiments. At one of the extremities of the façade of the Ecole Normale, on the same line as the doorkeeper's lodge, a pavilion had been built for the school architect and his clerk. Pasteur succeeded in obtaining possession of this small building, and transformed it into a laboratory. He built a drying stove under the staircase; though he could reach the stove only by crawling on his knees, this was better than his old attic. He also had a pleasant surprise—he was given a curator. He had deserved one sooner, for he had founded the institution of *agrégés préparateurs.* Remembering his own desire, on leaving the

Ecole Normale, to have a year or two for independent study, he had wished to facilitate for others the obtaining of those few years of research and perhaps inspiration. Thanks to him, five places as laboratory curators were exclusively reserved to Ecole Normale students who had taken their degree *(agrégés)*. The first curator who entered the new laboratory was Jules Raulin, a young man with a clear and sagacious mind, a calm and tenacious character, loving difficulties for the sake of overcoming them.

Pasteur began by the microscopic study of atmospheric air. "If germs exist in atmosphere," he said, "could they not be arrested on their way?" It then occurred to him to draw —through an aspirator—a current of outside air through a tube containing a little plug of cotton wool. The current as it passed deposited on this sort of filter some of the solid corpuscles contained in the air; the cotton wool often became black with those various kinds of dust. Pasteur assured himself that amongst various detritus those dusts presented spores and germs. "There are therefore in the air some organized corpuscles. Are they germs capable of vegetable productions, or of infusions? That is the question to solve." He undertook a series of experiments to demonstrate that the most putrescible liquid remained pure indefinitely if placed out of the reach of atmospheric dusts. But it was sufficient to place in a pure liquid a particle of the cotton-wool filter to obtain an immediate alteration.

A year before starting any discussion Pasteur wrote to Pouchet that the results which he had attained were "not founded on facts of a faultless exactitude. I think you are wrong, not in believing in spontaneous generation (for it is difficult in such a case not to have a preconceived idea), but in affirming the existence of spontaneous generation. In experimental science it is always a mistake not to doubt when facts do not compel affirmation. . . . In my opinion, the question is whole and untouched by decisive proofs. What is there in air which provokes organization? Are they germs? is it a solid? is it a gas? is it a fluid? is it a principle such as ozone? All this is unknown and invites experiment."

After a year's study, Pasteur reached this conclusion: "Gases, fluids, electricity, magnetism, ozone, things known or things occult, there is nothing in the air that is conditional to life, except the germs that it carries."

Pouchet defended himself vigorously. To suppose that germs came from air seemed to him impossible. How many millions of loose eggs or spores would then be contained in a cubic millimetre of atmospheric air?

"What will be the outcome of this giant's struggle?" grandiloquently wrote an editor of the *Moniteur Scientifique* (April, 1860). Pouchet answered this anonymous writer by advising him to accept the doctrine of spontaneous generation adopted of old by so many "men of genius." Pouchet's principal disciple was a lover of science and of letters, M. Nicolas Joly, an *agrégé* of natural science, doctor of medicine, and professor of physiology at Toulouse. He himself had a pupil, Charles Musset, who was preparing a thesis for his doctor's degree under the title: *New Experimental Researches on Heterogenia, or Spontaneous Generation.*

Pasteur intended to narrow more and more the sphere of discussion. It was an ingenious operation to take the dusts from a cotton-wool filter, to disseminate them in a liquid, and thus to determine the alteration of that liquid; but the cotton-wool itself was an organic substance and might be suspected. He therefore substituted for the cotton-wool a plug of asbestos fiber, a mineral substance. He invented little glass flasks with a long curved neck; he filled them with an alterable liquid, which he deprived of germs by ebullition; the flask was in communication with the outer air through its curved tube, but the atmospheric germs were deposited in the curve of the neck without reaching the liquid; in order that alteration should take place, the vessel had to be inclined until the point where the liquid reached the dusts in the neck.

But Pouchet said, "How could germs contained in the air be numerous enough to develop in every organic infusion? Such a crowd of them would produce a thick mist as dense as iron." Of all the difficulties this last seemed to Pasteur the hardest to solve. Could it not be that the dissemination of germs was more or less thick according to places? "Then," cried the heterogenists, "there would be sterile zones and fecund zones, a most convenient hypothesis, indeed!" Pasteur let them laugh whilst he was preparing a series of flasks reserved for divers experiments. If spontaneous generation existed, it should invariably occur in vessels filled with the same alterable liquid. "Yet it is ever possible," affirmed Pasteur, "to take up in certain places a notable though limited volume of ordinary air, having been submitted to no physical or chemical change, and still absolutely incapable of producing any alteration in an eminently putrescible liquor." He was ready to prove that nothing was easier than to increase or to reduce the number either of the vessels where productions should appear or of the vessels where those productions should be lacking. After introducing into a series of flasks of a capacity of 250 cubic centimeters a very easily corrupted liquid, such as yeast water, he submitted each flask to ebullition. The neck of those vessels was ended off in a vertical point. Whilst the liquid was still boiling, he closed, with an enameler's lamp, the pointed opening through which the steam had rushed out, taking with it all the air contained in the vessel. Those flasks were indeed calculated to satisfy both partisans or adversaries of spontaneous generation. If the extremity of the neck of one of these vessels was suddenly broken, all the ambient air rushed into the flask, bringing in all the suspended dusts; the bulb was closed again at once with the assistance of a jet of flame. Pasteur could then carry it away and place it in a temperature of 25–30° C., quite suitable for the development of germs and mucors.

In those series of tests some flasks showed some alteration, others remained pure, according to the place where the air had been admitted. During the beginning of the year 1860 Pasteur broke his bulb points and enclosed ordinary air in many different places, including the cellars of the Observatory of Paris. There, in that zone of an invariable temperature, the absolutely calm air could not be compared to the air he gathered in the yard of the same building. The results were also very different: out of ten vessels opened in the cellar, closed again and placed in the stove, only one showed any alteration; whilst eleven others, opened in the yard, all yielded organized bodies.

When the long vacation approached, Pasteur, who intended to go on a voyage of experiments, laid in a store of glass flasks. He wrote to Chappuis, on August 10, 1860: "I fear from your letter that you will not go to the Alps this year. . . . Besides the pleasure of having you for a guide, I had hoped to utilize your love of science by offering you the modest part of curator. It is by some study of air on heights afar from habitations and vegetation that I want to conclude my work on so-called spontaneous generation. The real interest of that work for me lies in the connection of this subject with that of ferments which I shall take up again November."

Pasteur started for Arbois, taking with him seventy-three flasks; he opened twenty of them not very far from his father's tannery, on the road to Dôle, along an old road, now a path which leads to the mount of the Bergère. The vine laborers who passed him wondered what this holiday tourist could be doing with all those little phials; no one suspected that he was penetrating one of nature's greatest secrets. "What would you have?" merrily said his old friend, Jules Vercel; "it amuses him!" Of those twenty vessels, opened some distance away from any dwelling, eight yielded organized bodies.

Pasteur went on to Salins and climbed Mount Poupet, 850 meters above the sea level. Out of twenty vessels opened, only five were altered. Pasteur would have liked to charter a balloon in order to prove that the higher you go the fewer germs you find, and that certain zones, absolutely pure, contain none at all. It was easier to go into the Alps.

He arrived at Chamonix on September 20, and engaged a guide to make the ascent of the Montanvert. The very next morning this novel sort of expedition started. A mule carried the case of thirty-three vessels, followed very closely by Pasteur, who watched over the precious burden and walked alongside of precipices supporting the case with one hand so that it should not be shaken.

When the first experiments were started an incident occurred. Pasteur has himself related this fact in his report to the Académie. "In order to close again the point of the flasks after taking in the air, I had taken with me an eolipyle spirit lamp. The dazzling whiteness of the ice in the sunlight was such that it was impossible to distinguish the jet of burning alcohol, and as moreover that was slightly moved by the wind, it never remained on the broken glass long enough to hermetically seal my vessel. All the means I might have employed to make the flame visible and consequently directable would inevitably have given rise to causes of error by spreading strange dusts into the air. I was therefore obliged to bring back to the little inn of Montanvert, unsealed, the flasks which I had opened on the glacier."

The inn was a sort of hut, letting in wind and rain. The thirteen open vessels were exposed to all the dusts in the room where Pasteur slept; nearly all of them presented alterations.

In the meanwhile the guide was sent to Chamonix, where a tinker undertook to modify the lamp in view of the coming experiment.

The next morning, twenty flasks, which have remained celebrated in the world of scientific investigators, were brought to the Mer de Glace. Pasteur gathered the air with infinite precaution; he used to enjoy relating these details to those people who call everything easy. After tracing with a steel point a line on the glass, careful lest dusts should become a cause of error, he began by heating the neck and fine point of the bulb in the flame of the little spiritlamp. Then raising the vessel above his head, he broke the point with steel nippers, the long ends of which had also been heated in order to burn the dusts which might be on their surface and which would have been driven into the vessel by the quick inrush of the air. Of those twenty flasks, closed again immediately, only one was altered. "If all the results are compared that I have obtained until now," he wrote, on March 5, 1880, when relating this journey to the Académie, "it seems to me that it can be affirmed that the dusts suspended in atmospheric air are the exclusive origin, the necessary condition of life in infusions."

And in an unnoticed little sentence, pointing already then to the goal he had in view, "What would be most desirable would be to push those studies far enough to prepare the road for a serious research into the origin of various diseases." The action of those little beings, agents not only of fermentation but also of disorganization and putrefaction, already dawned upon him.

While Pasteur was going from the Observatoire cellars to the Mer de Glace, Pouchet was gathering air on the plains of Sicily, making experiments on Etna, and on the sea. He saw everywhere, he wrote, "air equally favorable to organic genesis, whether surcharged with detritus in the midst of our populous cities, or taken on the summit of a mountain, or on the sea, where it offers extreme purity. With a cubic decimeter of air, taken where you like, I affirm that you can ever produce legions of microzoa."

And the heterogenists proclaimed in unison that "everywhere, strictly everywhere, air is constantly favorable to life." Those who followed the debate nearly all leaned toward Pouchet. "I am afraid," wrote a scientific journalist in *La Presse* (1860), "that the experiments you quote, M. Pasteur, will turn against you. . . . The world into which you wish to take us is really too fantastic. . . ."

The three heterogenists, Pouchet, Joly, and Musset, proposed to use that same time in fighting Pasteur on his own ground. They started from Bagnères-de-Luchon followed by several guides and taking with them all kinds of provisions and some little glass flasks with a slender pointed neck. They crossed the pass of Venasque without incident and decided to go further, to the Rencluse. Some isard-stalkers having come toward the strange-looking party, they were signaled away; even the guides were invited to stand aside. It was necessary to prevent any dusts from reaching the bulbs, which were thus opened at 8 p.m. at a height of 2,083 meters. But eighty-three meters higher than the Montanvert did not seem to them enough; they wished to go higher. "We shall sleep on the mountain," said the three scientists. Fatigued and bitter cold, they withstood everything with the courage inspired by a problem to solve. The next morning they climbed across that rocky chaos, and at last reached the foot of one of the greatest glaciers of the Maladetta, 3,000 meters above the sea level. "A very deep narrow crevasse," says Pouchet, "seemed to us the most suitable place for our experiments." Four phials (filled with a decoction of hay) were opened and sealed again with precautions that Pouchet considered as exaggerated.

Pouchet, in his merely scientific report, does not relate the return journey, yet more perilous than the ascent. At one of the most dangerous places, Joly slipped and would have rolled into a precipice but for the strength and presence of mind of one of the guides. All three at last came back to Luchon, forgetful of dangers run, and glorying at having reached 1,000 meters higher than Pasteur. They triumphed when they saw alteration in their flasks! "Therefore," said Pouchet, "the air of the Maladetta, and of high mountains in general, is not incapable of producing alteration in an eminently putrescible liquor; therefore heterogenia, or the production of a new being devoid of parents, but formed at the expense of ambient organic matter, is for us a reality."

The Academy of Sciences was taking more and more interest in this debate. In November, 1863, Joly and Musset expressed a wish that the Academy should appoint a Commission, before whom the principal experiments of Pasteur and of his adversaries should be repeated.

Already in the preceding year, the Académie itself had evidenced its opinion by giving Pasteur the prize of a competition proposed in these terms: "To attempt to throw some new light upon the question of so-called spontaneous generation by well-conducted experiments." Pasteur's treatise on *Organized Corpuscles Existing in Atmosphere* had been unanimously preferred. Pasteur might have entrenched himself behind the suffrages of the Academy, but begged it, in order to close those incessant debates, to appoint the Commission demanded by Joly and Musset.

The members of the Commission were Flourens, Dumas, Brongniart, Milne-Edwards, and Balard. Pasteur wished that the discussion should take place as soon as possible, and it was fixed for the first fortnight in March. But Pouchet, Joly, and Musset asked for a delay on account of the cold. "We consider that it might compromise, perhaps prevent, our results, to operate in a temperature which often goes below zero even in the south of France. How do we know that it will not freeze in Paris between the first and fifteenth of March?" They even asked the Commission to adjourn experiments until the summer. "I am much surprised," wrote Pasteur, "at the delay sought by Messrs. Pouchet, Joly, and Musset; it would have been easy with a stove to raise the temperature to the degree required by those gentlemen. For my part I hasten to assure the Academy that I am at its disposal, and that in summer, or in any other season, I am ready to repeat my experiments."

Some evening scientific lectures had just been inaugurated at the Sorbonne; such a subject as spontaneous generation was naturally on the program. When Pasteur entered the large lecture room of the Sorbonne on April 7, 1864, he must have been reminded of the days of his youth, when crowds came, as to a theatrical performance, to hear J. B. Dumas speak. Dumas' pupil, now a master, in his turn found a still greater crowd invading every corner. Amongst the professors and students, such celebrities as Duruy, Alexandre Dumas senior, George Sand, Princess Mathilde, were being pointed out. Around them, the inevitable "smart" people who must see everything and be seen everywhere, without whom no function favored by fashion would be complete; in short what is known as the "Tout Paris." But this "Tout Paris" was about to receive a novel impression, probably a lasting one. The man who stood before this fashionable audience was not one of those speakers who attempt by an insinuating exordium to gain the good graces of their hearers; it was a grave-looking man, his face full of quiet energy and reflective force.

With perfect clearness and simplicity, Pasteur explained how the dusts which are suspended in air contain germs of inferior organized beings and how a liquid preserved, by certain precautions, from the contact of these germs can be kept indefinitely, giving his audience a glimpse of his laboratory methods.

"Here," he said, "is an infusion of organic matter, as limpid as distilled water, and extremely alterable. It has been prepared today. Tomorrow it will contain animalculae, little infusories, or flakes of moldiness.

"I place a portion of that infusion into a flask with a long neck, like this one. Suppose I boil the liquid and leave it to cool. After a few days, moldiness or animalculae will develop in the liquid. By boiling, I destroyed any germs contained in the liquid or against the glass; but that infusion being again in contact with air, it becomes altered, as all infusions do. Now suppose I repeat this experiment, but that, before boiling the liquid, I draw (by means of an enameler's lamp) the neck of the flask into a point, leaving, however, its extremity open. This being done, I boil the liquid in the flask, and leave it to cool. Now the liquid of this second flask will remain pure not only two days, a month, a year, but three or four years—for the experiment I am telling you about is already four years old, and the liquid remains as limpid as distilled water. What difference is there, then, between those two vases? They contain the same liquid, they both contain air, both are open! Why does one decay and the other remain pure? The only difference between them is this: in the first case, the dusts suspended in air and their germs can fall into the neck of the flask and arrive into contact with the liquid, where they find appropriate food and develop; thence microscopic beings. In the second flask, on the contrary, it is impossible, or at least extremely difficult, unless air is violently shaken, that dusts suspended in air should enter the vase; they fall on its curved neck. When air goes in and out of the vase through diffusions or variations of temperature, the latter never being sudden, the air comes in slowly enough to drop the dusts and germs that it carries at the opening of the neck or in the first curves.

"This experiment is full of instruction; for this must be noted, that everything in air save its dusts can easily enter the vase and come into contact with the liquid. Imagine what you choose in the air—electricity, magnetism, ozone, unknown forces even, all can reach the infusion. Only one thing cannot enter easily, and that is dust, suspended in air. And the proof of this is that if I shake the vase violently two or three times, in a few days it contains animalculae or moldiness. Why? because air has come in violently enough to carry dust with it.

"And, therefore, gentlemen, I could point to that liquid and say to you, I have taken my drop of water from the immensity of creation, and I have taken it full of the elements appropriated to the development of inferior beings. And I wait, I watch, I question it, begging it to recommence for me the beautiful spectacle of the first creation. But it is dumb, dumb since these experiments were begun several years ago; it is dumb because I have kept it from the only thing man cannot produce, from the germs which float in the air, from Life, for Life is a germ and a germ is Life. Never will the doctrine of spontaneous generation recover from the mortal blow of this simple experiment."

The public enthusiastically applauded these words, which ended the lecture:

"No, there is now no circumstance known in which it can be affirmed that microscopic beings came into the world without germs, without parents similar to themselves. Those who affirm it have been duped by illusions, by ill-conducted experiments, spoilt by errors that they either did not perceive or did not know how to avoid."

In the meanwhile, besides public lectures and new studies, Pasteur succeeded in "administering" the Ecole Normale in the most complete sense of the word. His influence was such that students acquired not a taste but a passion for study; he directed each one in his own line, he awakened their instincts. It was already through his wise inspiration that five "Normaliens agrégés" should have the chance of the five curators' places; but his solicitude did not stop there. If some disappointment befell some former pupil, still in that period of youth which doubts nothing or nobody, he came vigorously to his assistance; he was the counselor of the future. A few letters will show how he understood his responsibility.

A Normalien, Paul Dalimier, received 1st at the *agrégation* of Physics in 1858, afterwards Natural History curator at the Ecole, and who, having taken his doctor's degree, asked to be sent to a Faculty, was ordered to go to the Lycée of Chaumont.

In the face of this almost disgrace he wrote a despairing letter to Pasteur. He could do nothing more, he said, his career was ruined. "My dear sir," answered Pasteur, "I much regret that I could not see you before your departure for Chaumont. But here is the advice which I feel will be useful to you. Do not manifest your just displeasure; but attract attention from the very first by your zeal and talent. In a word, aggravate, by your fine discharge of your new duties, the injustice which has been committed. The discouragement expressed in your last letter is not worthy of a man of science. Keep but three objects before your eyes: your class, your pupils, and the work you have begun. . . . Do your duty to the best of your ability without troubling about the rest."

Pasteur undertook the rest himself. He went to the Ministry to complain of the injustices and unfairness, from a general point of view, of that nomination.

"Sir," answered the Chaumont exile, "I have received your kind letter. My deep respect for every word of yours will guarantee my intention to follow your advice. I have given myself up entirely to my class. I have found here a Physics cabinet in a deplorable state, and I have undertaken to reorganize it."

He had not time to finish: justice was done and Paul Dalimier was made *maître des conférences* at the Ecole Normale. He died at twenty-eight.

At that same time, the heterogenists had at last placed themselves at the disposal of the Académie and were invited to meet Pasteur before the Natural History Commission at M. Chevreul's laboratory. "I affirm," said Pasteur, "that in any place it is possible to take up from the ambient atmosphere a determined volume of air containing neither egg nor spore and producing no generation in putrescible solutions." The Commission declared that, the whole content bearing upon one simple fact, one experiment only should take place. The heterogenists wanted to recommence a whole series of experiments, thus reopening the discussion. The Commission refused, and the heterogenists, unwilling to concede the point, retired from the field, repudiating the arbiters that they had themselves chosen.

And yet Joly had written to the Académie, "If one only of our flasks remains pure, we will loyally own our defeat." A scientist who later became Permanent Secretary of the Académie des Sciences, Jamin, wrote about this conflict:

"The heterogenists, however they may have colored their retreat, have condemned themselves. If they had been sure of the fact—which they had solemnly engaged to prove or to own themselves vanquished—they would have insisted on showing it, it would have been the triumph of their doctrine."

With Pasteur's masterly experiments one aspect of the problem of the origin of life seemed definitely decided: life could originate only from life already in existence. To those who insisted that this conclusion merely begged the question, that "if all comes from a germ, whence came the first germ?" Pasteur merely replied, "We must bow before that question." In making that statement he failed to foresee twentieth-century developments. It appears from the quotation in the following article that Darwin permitted himself to speculate on these developments in a surprisingly accurate manner. The question of "whence came the first germ" has been removed from the area of philosophy and religion to that of experimental science. The theories and techniques involved are described in the following article. Born in Hungary, Ernest Borek became a Ph.D. at Columbia in 1938. He has been a faculty member at Columbia, a scholar at the Pasteur Institute and a consultant of the Office of Scientific Research and Development of the United States Government. He is now Professor of Chemistry at the College of the City of New York.

ATOMS INTO LIFE

ERNEST BOREK

APART FROM Divine Creation, there have been two different suggestions on the origin of life on Earth. The first one was proposed, among others, by the distinguished chemist Arrhenius, whose enduring contribution was the recognition that matter can exist in the form of electrically charged particles. He coined the term *ions* for such particles. In 1908 Arrhenius suggested that life on Earth started from some spores which drifted here from interplanetary space.

The propelling force for the interplanetary or interstellar space voyages was supposed to be the pressure exerted by light rays. Arrhenius calculated that a bacterial spore could move with great speed in the interstellar voids. He estimated that one could reach us from Alpha Centauri in 9,000 years. Such a suggestion, of course, evades the issue: it is nothing more than buck-passing on an astronomical scale. Moreover, we now know from our rocket probings of outer space that Earth is surrounded by a zone rich in ions of high energy, the so-called Van Allen belt. The passage of a spore in a viable state through this zone, unprotected by a lead shield, is unlikely. The Arrhenius theory on the origin of life is thus riddled by the very particles he named. Moreover, it is difficult to see how the spores, without ceramic nose cones, could survive the frictional heat once they were in contact with the atmosphere.

Of the indigenous origin of life on Earth there are two different theories that need to be seriously considered: The first one holds that the primordial living forms were autotrophic, that is, they could synthesize all of the components of their structure from the inorganic substances available on the surface of Earth, just as our contemporary green plants do.

Such a living form would have to arise, equipped with arsenals of integrated enzymes to perform the intricate tasks of weaving atoms into substances which theretofore did not exist. The probability of the chance occurrence of such an organism is so small that this theory in effect invokes Divine Creation. From the point of view of an experimental scientist, a theory which depends on a miraculous event which occurred eons ago is sterile: It is beyond our experimental reach, there is nothing we can do with it beyond stating it.

The second theory postulates that the primordial forms were merely accretions of complex organic compounds which we presume abounded in the oceans of the primitive Earth. These preliving forms merely agglomerated the many compounds in the laden seas, and only eons later did they acquire synthetic abilities. According to this hypothesis the autotrophs, organisms with complete synthetic capacity, were the result of progressive chemical evolution. This conjecture is the most persuasive for the experimental biochemist.

The years 1857 to 1859 were bountiful years in our cultural history. Pasteur published his incisive studies on fermentation in 1857. This not only led to a refutation of the false theory of spontaneous generation, but also started us on the path leading to an understanding of the mechanism of life as an integrated series of physicochemical reactions catalyzed by enzymes. In 1858 and 1859 Wallace and Darwin published the theory of evolution by natural selection. This brilliant insight into the workings of nature suddenly revealed the road which the complex modern creatures, be it man or mole, must have traveled in their ascent from some primordial living prototype. In the hundred years since, biochemists have accumulated irrefutable evidence at the molecular level for the common origin of contemporary creatures: The mighty whale swings his tail with the aid of adenosine triphosphate; the same compound of intricate structure also serves as life's battery for the quaint little microorganism *Thiobacillus thiooxidans,* which makes a living out of oxidizing sulfur to sulfuric acid. The chance occurrence of this and of other compounds for the same function in these two organisms without assuming common ancestry is of the same order of improbability as finding on a distant planet or another solar system a copy of the Taj Mahal identical to the last stone and also housing the remains of some potentate.

The theory, or better, the law of evolution through natural selection eliminated the need to consider the origin of life for each species. But it insistently forced on our minds the problem of the origin of the first ancestral living forms; for the law of evolution is a law of dynamic change, not of static origin.

The idea of the spontaneous generation of some prototype living form from the accumulated chemicals on the cooling primitive Earth became part of the *Zeitgeist* of the latter part of the nineteenth century. The German philosopher Ernst Haeckel wrote in 1868 in *The History of Creation* that the first living or pseudo-living forms must have been merely "homogeneous, structureless, amorphous lumps of protein" originating from dissolved matter in the primeval seas. That Darwin himself entertained such ideas is obvious from a letter he wrote in 1871. In it he defended the idea of such primeval creation against the objection of the lack of repetition of it at the present time.

> It is often said [he wrote] that all the conditions for the first production of a living organism are now present, which could ever have been present. But if (and oh! what a big if!) we could conceive in a warm little pond, with all sorts of ammonia and phosphoric salts, light, heat, electricity, etc., present, that a protein was chemically formed ready to undergo still more complex changes, at the present day such matter would be instantly devoured or absorbed, which would not have been the case before living creatures were formed.

It is interesting that both Haeckel and Darwin speak of proteins as the key structure of their primeval forms some twenty years before enzymes were discovered and almost sixty years before enzymes were proved to be proteins.

Implicit in this theory of the agglomeration of chemicals to form protein is the presence in the primeval seas of the component parts of proteins, the amino acids. There were several attempts to test the possible synthesis of amino acids under the conditions which may have existed on the primitive Earth. Mixtures of gases, presumably simulating those that abounded on the cooling Earth, were subjected to electric sparking or to ultraviolet irradiation and were analyzed for the presence of sugars and of amino acids. The results were unimpressive. The analytical methods were so poor until recently that the claims for the synthesis of substances other than sugars under the conditions reported were not convincing. Moreover, all of the investigators made the wrong assumption, that carbon had been present in the primitive Earth's atmosphere in the form of carbon dioxide.

In 1936 a Russian biochemist, Academician A. I. Oparin, published a fascinating book, *The Origin of Life*. This has since become the source book for any discussion in this area which tries to be rooted not in fantasy but in the few available facts.

In the first place, Oparin concluded that carbon must have been present in the primitive atmosphere of Earth, not as carbon dioxide but in the form of hydrocarbons. These are compounds made of chains of carbon festooned with hydrogen atoms. Hydrocarbons can be the sources for a large variety of different organic compounds—alcohols, acids, aldehydes—some of which could react with ammonia to form amino acids.

But a still more plausible nonliving source of amino acids was soon to be demonstrated. Dr. Harold C. Urey, on returning to academic life after the Second World War, left the field of isotopes and took up, of all things, the study of the origin of the planets. He has since made contributions equally as brilliant as his earlier works on the concentration of isotopes. Dr. Urey, though completely unaware of Oparin's work, calculated that the stable form of carbon in the presence of excess hydrogen on the primitive Earth would be methane. This is the most primitive hydrocarbon, consisting of one atom of carbon to which are attached four atoms of hydrogen. Similarly, nitrogen would be most likely present in the form of ammonia, which is composed of an atom of nitrogen with three atoms of hydrogen joined to it. This too was a reaffirmation of Oparin's conclusion.

Confirmation of the Oparin-Urey hypothesis on the nature of the primitive atmosphere of the planets is provided by astrophysics. From a study of their spectra we are certain that the atmosphere of the giant planets Jupiter, Saturn, and Uranus still consist of methane and ammonia. The reasons for their primitive state are threefold. Their temperature is much lower than of Earth, and therefore the chemical reactions are slower; since they are far away from the sun, the ultraviolet energy falling on them is less intense; and, finally, since their mass is greater, their gravitational pull is greater and thus they could keep their original atmosphere captive for a longer time.

If methane rather than carbon dioxide was the primeval link from which the chains of organic molecules were forged, then the reason for the failure to produce amino acids in attempts to synthesize these compounds under the erroneously presumed conditions becomes clear. Even the best chef's soufflé will not rise if he uses the yolk instead of the white of the egg.

Dr. Urey wrote in 1952:

> It seems to me that experimentation on the production of organic compounds from water and methane in the presence of ultraviolet light of approximately the spectral distribution estimated for sunlight would be most profitable. The investigation of possible effects of electric discharges on the reaction should also be tried since electric storms in the reducing atmosphere can be postulated reasonably.

Such experiments indeed did prove profitable. The passing of electric sparks through a reconstructed primitive Earth atmosphere was worked out by one of Dr. Urey's young graduate students, Dr. Stanley L. Miller. Such an experiment was carried out by him somewhat as follows. Into an appropriately constructed instrument were introduced water and the three gases, methane, ammonia, and hydrogen. The glass apparatus was so designed that the water could be boiled to a vapor, then condensed to a liquid, and electric sparks could be continuously passed through the area where water vapor and the three gases were passing. The cooling and sparking in the absence of oxygen was continued for as long as a week. At the end of this time the reaction was stopped, the vessel was cooled, and it was opened to allow the gases to escape. Then, after the water was evaporated, came the exciting moment. There was a solid residue. If there were no reaction there would have been no residue, for initially there were no solids in the system.

The analysis of the newly created solids was a very simple procedure by paper chromatography, the ingenious method for the analysis of minute amounts of chemicals. The solids turned out to be a mixture of a variety of organic compounds, including several amino acids in large amounts. This was exciting news, but more recently even more satisfying findings were reported by another young biochemist. If to the same mixture of gases some iron sulfide is added, and it is then irradiated by ultraviolet light, some of the other amino acids such as phenylalanine and the sulfur-containing methionine are produced, as well as complex aggregates of these and other amino acids. The synthesis of these complexes, or polypeptides, is of particular interest, for they are held together by the same binding system which builds protein molecules. Since polypeptides are assembled this way, the weaving of still more amino acids into a larger protein molecule may be just a matter of time. Any system which can weave a napkin can also turn out a tablecloth.

These are extraordinarily suggestive discoveries. The structure of phenylalanine is rather intricate. It consists of a ring of six carbon atoms, to five of which single hydrogen atoms are attached. From the sixth carbon stems an appendage of three carbon atoms. To the second of these carbons is attached a nitrogen plus two hydrogen atoms, or the amino group; to the third carbon are attached two oxygens and a hydrogen, the acidic group.

The alignment of these twenty-three atoms of four different atomic species into the exact pattern of one of our essential amino acids under the influence of ultraviolet irradiation appears to be almost magical. And indeed it is: the magic of molecular reactivity. To the chemist, however, such molecular transformations induced by heat, pressure, and irradiations are less than magical; they are the source of his daily bread. The most important step in making that wonder drug aspirin is to heat together, under pressure, carbolic acid and carbon dioxide—presto! out comes salicylic acid.

How is that particular substance, phenylalanine, formed under the influence of ultraviolet light? The fact is that there are literally hundreds of other compounds formed under those conditions. But most of these are too unstable to exist for more than fractions of a second. Phenylalanine and the other amino acids are stable, condensed systems of the atoms involved, and once they are formed they endure, we know from work on fossils, for as long as three hundred million years. Thus they will accumulate at the expense of the other unstable transient forms.

A surprising aspect of the amino acids made under these conditions is that they are almost uniformly the ones with the amino groups on the second carbon atom. These are the so-called *alpha* amino acids. They are the ones our body's proteins, as well as the proteins of fossils, are made of. Had the tendency of carbon atoms been different, say to put the amino group on the third carbon, that is, to form *beta* amino acids, our proteins would have had entirely different properties; we would probably be entirely different looking creatures. But this is only one aspect of the life-shaping power of the carbon atom. The promise of life itself was locked into the structure of the carbon atom the moment it came into being. For, of all of the elements known, carbon has the greatest capacity for uniting with itself into molecules of vast size and infinite variety. Only from so versatile an element, which has the ability to rearrange itself into an infinity of patterns, each with new and unique properties, could so wondrously complex a system as life arise. It has been suggested that life was created the moment matter was created. This may be true in the sense that the spark of life started flickering in the valence electrons of the carbon atom.

How did life arise? I ought to warn the reader that the only facts known are those I have outlined above: amino acids and other organic molecules could be formed with ease on the primitive Earth. The next epoch we can document is the Cambrian period, the time of the oldest remnant fossils. What happened in the billions of years in between is, at the present time, almost anybody's guess. Conjecturing on it is a game everybody can play and, indeed, almost everybody does. It is a pleasant game for many of us. A career in science is often a career of intense preoccupation with minutiae. It is a joy from time to time to escape our narrow corner, cast off the restrictions imposed on us by the methods and the ethics of science, not to mention the editors of science, and assume the role of cosmologists. Speculation on events which occurred eons ago under conditions which cannot be repeated is an engaging pastime, one that is nowhere nearly so taxing as establishing a tiny little fact about our contemporary world which may be promptly challenged by a contemporary colleague.

If the reader would like to come and have coffee with me, here's the way one speculative monologue on the origin of life might go.

Let us visualize the face of the primitive Earth. There are many lines of evidence indicating that at one time it was a seething hot mass which slowly cooled to a temperature where the enveloping clouds of steam could condense to form the vast seas. The atmosphere, say about three billion years ago, consisted essentially of methane, ammonia, hydrogen, and water vapor. Enormous amounts of energy cascaded on this mixture of gases in the form of sunlight (at the present time this rate is 260,000 calories per square centimeter per year). Lightning, from electrical storms, also provided energy, albeit most likely at a much lower level. (At the present time this rate is only about 4 calories per square centimeter per year.)

Prodigious amounts of organic compounds were continuously being produced in the atmosphere. As fast as they were shaped, these compounds cascaded below with the frequent torrential rains to accumulate in the seas below. If all the carbon which is estimated to be present on the surface of Earth today were in such soluble organic form, the seas might have been laden with two billion billion (2×10^{18}) tons of material. J. B. S. Haldane, the English biologist, who is one of those scientists who speculates in print on the origin of life, called these compound-laden seas a "hot thin soup." If all the carbon was really present in the form of organic compounds, especially as polypeptides, it was quite a "thick soup." (These speculations do not deserve more precise units of measurement than "thick" and "thin.")

Whatever one's preference for soup, be it bouillion or mulligatawny, the primeval seas undoubtedly abounded in the organic compounds which could serve as the building blocks of eventual life. What particular substances may have been present in those primeval seas?

It is certain that the shaping of organic molecules in the primitive atmosphere was far more prolific than would be indicated by the few experiments which have been performed to probe this area. With a positive finding we are twice blessed: we have not only something new, but a rich promise of more findings to come; with a negative finding we are twice vexed: a reaction may just not have gone under one set of conditions, or it may never go. If a chemist reports that he found no evidence for the synthesis of purines or porphyrins after exposing the presumed primordial gases to sparking or to irradiations for a few days, he is really saying that under one set of conditions he has not found them in concentrations high enough for the assay methods he used. Our most sensitive tests will not detect these substances in amounts below a ten-millionth of a gram (10^{-7} g.). The number of molecules which can still elude the prying chemist is astronomical. It is on the order of a hundred thousand billion (10^{14}) molecules.

The chemist runs his experiments for days, in tiny volumes. What might have accumulated in the primitive oceans in a few hundred million years is beyond our imaginings. Some of the silicate mineral beds along the shores could have selectively concentrated the rarest of organic molecules.

It is probably significant that the structural units which are needed for the assembly of life are extraordinarily rugged compounds. The stability of the amino acids has already been discussed. Equally rugged are the porphyrins. These molecules serve as the cutting edge of several pivotal catalytic molecules: they are present in chlorophyll, where they are involved in photosynthesis; in hemoglobin they transport oxygen; in the cytochromes they perform cellular respiration. Coproporphyrin is an excreted end product of porphyrins resulting from the death of red cells. Coproporphyrin is so stable it has been found intact in the fossilized excreta of crocodiles which may have been basking in the sun 50 million years ago. The purines and pyrimidines which are the building blocks for the nucleic acids are also very stable molecular structures. It would appear that life evolved out of molecules which could sit around in the primeval seas for years, waiting to be picked up, without becoming shopworn.

How then did life arise from this fluid warehouse of stable organic molecules? We are under a profound handicap when we come to grips with this problem: we have no models to look at! The biological scientist can but observe life's mechanism, he can rarely predict them. The one certainty that emerges from a study of the history of the biological sciences is that nature's methods have invariably outstripped our imagination. Could the human mind have conceived that we have ascended from some notochord-bearing little fish like a lancelet or an eel, without the evidence in the rocks staring us in the face? Could we have visualized a priori the living machine driven on phosphate bond energy?

Unfortunately, there will be no models of pre-living forms to look at until possibly some day a well-trained biochemist may be landed on Mars. And, alas, we may be too late even there, for that planet's oceans have evaporated. If the changes in color with the changing seasons indicate vegetation, as they do to some experts, then Mars has quite an advanced form of life.[1]

Meanwhile we must be satisfied with conjectures, for we cannot dignify our speculations by assigning to them the well-defined term in the semantics of science, hypothesis. A hypothesis should be an integration of observed phenomena, and it must predict other phenomena, either existing or wrought by further experiments. How can we design an experiment to repeat an event which may have been the culmination of perhaps a billion years of molecular interactions?

1. Observations made by instruments carried by a Mariner rocket after this article was written indicate that advanced life may not exist on Mars.—Eds.

Every conjecture on the origin of life which has been put forth is essentially an extrapolation of some of the known mechanisms in contemporary living cells. We merely endow some primordial protein molecule which was shaped through the inexorable laws of permutation and combination with some of the known properties of today's protein molecules. For example, we speculate that the photosynthetic mechanism could fumblingly start when some clump of protein molecules, endowed by their random structure with catalytic properties, was tossed by a wave onto a magnesium silicate deposit encrusted with porphyrin molecules. For these are the essential ingredients of a contemporary photosynthetic system: porphyrin, magnesium, and protein enzymes. By extrapolation of some of the known laws of microbial genetics we ascribe a survival potential to such a primordial clump. Other protein clumps, without synthetic capacity, would exhaust some of their nutrients and would be doomed to disintegration, whereas the clumps with some synthetic capacity could survive a temporary local famine. The more synthetic potencies a clump would have the greater would be its chance of survival. Thus, natural selection could operate at the molecular level, and it would favor the survival of aggregates with increasingly complex synthetic capacity and structure.

Such speculations are stimulating, but, it appears to the writer, scientifically not fruitful; a speculation which can lead to no experiment which could increase our arsenal of knowledge is barren.

What can we know about these preliving forms which could leave no traces in the ancient seas? They may have been microscopic, or as large as contemporary jellyfish, or they may have been vast blobs the size of a continent. It is enough for the present to assume that just as under a given set of circumstances the most complex of amino acids are assembled from small prototype molecules with the energy of sunlight in a brief time, so in a billion years a living, self-duplicating form might have arisen by molecular evolution from the precursor building blocks abounding in the primeval seas. Just as an amino acid can accumulate because of its stability at the expense of less stable molecular configurations, so could such living forms be selected from the billions of random experiments of nature through their own unique aptitude for survival: the ability for self-duplication.

If such molecular evolution seems implausible, so does organic evolution, despite the evidence written in the rocks, seem implausible. It is easier for the writer to believe that living unicellular organisms arose after a billion years from random molecular experiments of nature than that Beethoven and Einstein arose from some gasping little eel.

A materialist theory of the origin of life from preformed molecules is persuasive because it assumes a gradual evolution starting at the molecular level. Evolutionary transformation is a pervasive pattern in our universe. Nature, that arch conservative, abhors abrupt change. The atoms have evolved from prototype particles, the lightest atoms were packed into heavier ones; the solar system repeated on a gigantic scale the structural pattern of the atom. Gradual evolution has certainly been the ruling pattern of life in the past couple of hundred million years. Molecular evolution may have preceded this by a couple of billion years.

It is only the first stage of this theory, the aggregation of random protein molecules into a self-perpetuating system, which is highly improbable. After that the theory, with its life-evolving progression, gains in probability.

But, inherent in every description of creation is an increasing improbability as ultimate origins are approached. The creation of Adam was a miracle second only to the miracle of the creation of the Universe. But the shaping of Eve out of a living rib was merely a bravura feat of tissue culture.

IV. Evolution and Genetics

IV. *Evolution and Genetics*

The greatest name in biological science is that of Charles Robert Darwin, the gentle, cautious, morally upright Englishman who, with Alfred Russel Wallace, is the author of the theory of evolution. Born at Shrewsbury, England, in 1809, the son of a physician and grandson of the poet, scientist, and physician, Erasmus Darwin, he first took up medical studies at Edinburgh and then studied for the church at Cambridge. He lacked an abiding interest in either pursuit, to the disappointment of his father, who once said to him, "You care for nothing but shooting, dogs and rat-catching and will be a disgrace to yourself and all your family." Never was a judgment more mistaken. The young man had a consuming interest in natural history, and through the botanist John Stevens Henslow secured an appointment as naturalist aboard a global surveying expedition of the H.M.S. Beagle. *Extending from 1831 to 1836, the voyage was one of the most portentous in human history, for it provided Darwin with the stimulus which led him to his great discovery. As a result of observations, particularly in the Galapagos, he began to question the doctrine of the immutability of species. The notes he took served as the basis for his book* On the Origin of Species by Means of Natural Selection. *It was published in 1859, nearly two decades after the* Beagle *had returned to England. The interim was occupied in the most detailed and painstaking examination of his own ideas, and in recurring doubt, hesitation, and delay. The following selections, taken from his* Autobiography *and from a notable biography by Geoffrey West entitled* Charles Darwin, *throws light on the character of this extraordinary genius and on the dramatic events which resulted in final publication of* The Origin of Species.

FROM THE AUTOBIOGRAPHY

CHARLES DARWIN

Voyage of the Beagle*: from December 27, 1831, to October 2, 1836*

ON RETURNING HOME from my short geological tour in North Wales, I found a letter from Henslow, informing me that Captain Fitz-Roy was willing to give up part of his own cabin to any young man who would volunteer to go with him without pay as naturalist to the Voyage of the *Beagle.* I have given, as I believe, in my MS. Journal an account of all the circumstances which then occurred; I will here only say that I was instantly eager to accept the offer, but my father strongly objected, adding the words, fortunate for me, "If you can find any man of common sense who advises you to go I will give my consent." So I wrote that evening and refused the offer. On the next morning I went to Maer to be ready for September 1st, and whilst out shooting, my uncle sent for me, offering to drive me over to Shrewsbury and talk with my father, as my uncle thought it would be wise in me to accept the offer. My father always maintained that [my uncle] was one of the most sensible men in the world, and he at once consented in the kindest manner. I had been rather extravagant at Cambridge, and to console my father, said, "that I should be deuced clever to spend more than my allowance whilst on board the *Beagle*"; but he answered with a smile, "But they tell me you are very clever."

Next day I started for Cambridge to see Henslow, and thence to London to see Fitz-Roy, and all was soon arranged. Afterwards, on becoming very intimate with Fitz-Roy, I heard that I had run a very narrow risk of being rejected on account of the shape of my nose! He was an ardent disciple of Lavater, and was convinced that he could judge of a man's character by the outline of his features; and he doubted whether any one with my nose could possess sufficient energy and determination for the voyage. But I think he was afterwards well satisfied that my nose had spoken falsely.

Fitz-Roy's character was a singular one, with very many noble features: he was devoted to his duty, generous to a fault, bold, determined, and indomitably energetic, and an ardent friend to all under his sway. He would undertake any sort of trouble to assist those whom he thought deserved assistance. He was a handsome man, strikingly like a gentleman, with highly courteous manners, which resembled those of his maternal uncle, the famous Lord Castlereagh, as I was told by the Minister at Rio. Nevertheless he must have inherited much in his appearance from Charles II, for Dr. Wallich gave me a collection of photographs which he had made, and I was struck with the resemblance of one to Fitz-Roy; and on looking at the name, I found in Ch. E. Sobieski Stuart, Count d'Albanie, a descendant of the same monarch.

Fitz-Roy's temper was a most unfortunate one. It was usually worst in the early morning, and with his eagle eye he could generally detect something amiss about the ship, and was then unsparing in his blame. He was very kind to me, but was a man very difficult to live with on the intimate terms which necessarily followed from our messing by ourselves in the same cabin. We had several quarrels; for instance, early in the voyage at Bahia, in Brazil, he defended and praised slavery, which I abominated, and told me that he had just visited a great slave owner, who had called up many of his slaves and asked them whether they were happy, and whether they wished to be free, and all answered "No." I then asked him, perhaps with a sneer, whether he thought that the answer of slaves in the presence of their master was worth anything? This made him excessively angry, and he said that as I doubted his word we could not live any longer together. I thought that I should have been compelled to leave the ship; but as soon as the news spread, which it did quickly, as the captain sent for the first lieutenant to assuage his anger by abusing me, I was deeply gratified by receiving an invitation from all the gun-room officers to mess with them. But after a few hours Fitz-Roy showed his usual magnanimity by sending an officer to me with an apology and a request that I would continue to live with him.

His character was in several respects one of the most noble which I have ever known.

The voyage of the *Beagle* has been by far the most important event in my life, and has determined my whole career; yet it depended on so small a circumstance as my uncle offering to drive me thirty miles to Shrewsbury, which few uncles would have done, and on such a trifle as the shape of my nose. I have always felt that I owe to the voyage the first real training or education of my mind; I was led to attend closely to several branches of natural history, and thus my powers of observation were improved, though they were always fairly developed.

The investigation of the geology of all the places visited was far more important, as reasoning here comes into play. On first examining a new district, nothing can appear more hopeless than the chaos of rocks; but by recording the stratification and nature of the rocks and fossils at many points, always reasoning and predicting what will be found elsewhere, light soon begins to dawn on the district, and the structure of the whole becomes more or less intelligible. I had brought with me the first volume of Lyell's *Principles of Geology,* which I studied attentively; and the book was of the highest service to me in many ways. The very first place which I examined, namely, St. Jago, in the Cape de Verde Islands, showed me clearly the wonderful superiority of Lyell's manner of treating geology, compared with that of any other author whose works I had with me or ever afterwards read.

Another of my occupations was collecting animals of all classes, briefly describing and roughly dissecting many of the marine ones; but from not being able to draw, and from not having sufficient anatomical knowledge, a great pile of MS. which I made during the voyage has proved almost useless. I thus lost much time, with the exception of that spent in acquiring some knowledge of the Crustaceans, as this was of service when in after years I undertook a monograph of the Cirripedia.

During some part of the day I wrote my Journal, and took much pains in describing carefully and vividly all that I had seen; and this was good practice. My Journal served also, in part, as letters to my home, and portions were sent to England whenever there was an opportunity.

The above various special studies were, however, of no importance compared with the habit of energetic industry and of concentrated attention to whatever I was engaged in, which I then acquired. Everything about which I thought or read was made to bear directly on what I had seen or was likely to see; and this habit of mind was continued during the five years of the voyage. I feel sure that it was this training which has enabled me to do whatever I have done in science.

Looking backwards, I can now perceive how my love for science gradually preponderated over every other taste. During the first two years my old passion for shooting survived in nearly full force, and I shot myself all the birds and animals for my collection; but gradually I gave up my gun more and more, and finally altogether, to my servant, as shooting interfered with my work, more especially with making out the geological structure of a country. I discovered, though unconsciously and insensibly, that the pleasure of observing and reasoning was a much higher one than that of skill and sport. That my mind became developed through my pursuits during the voyage is rendered probable by a remark made by my father, who was the most acute observer whom I ever saw, of a skeptical disposition, and far from being a believer in phrenology; for on first seeing me after the voyage, he turned round to my sisters, and exclaimed, "Why, the shape of his head is quite altered."

To return to the voyage. On September 11th (1831), I paid a flying visit with Fitz-Roy to the *Beagle* at Plymouth. Thence to Shrewsbury to wish my father and sisters a long farewell. On October 24th I took up my residence at Plymouth, and remained there until December 27th, when the *Beagle* finally left the shores of England for her circumnavigation of the world. We made two earlier attempts to sail, but were driven back each time by heavy gales. These two months at Plymouth were the most miserable which I ever spent, though I exerted myself in various ways. I was out of spirits at the thought of leaving all my family and friends for so long a time, and the weather seemed to me inexpressibly gloomy. I was also troubled with palpitation and pain about the heart, and like many a young ignorant man, especially one with a smattering of medical knowledge, was convinced that I had heart disease. I did not consult any doctor, as I fully expected to hear the verdict that I was not fit for the voyage, and I was resolved to go at all hazards.

I need not here refer to the events of the voyage—where we went and what we did—as I have given a sufficiently full account in my published Journal. The glories of the vegetation of the Tropics rise before my mind at the present time more vividly than anything else; though the sense of sublimity, which the great deserts of Patagonia and the forest-clad mountains of Tierra del Fuego excited in me, has left an indelible impression on my mind. The sight of a naked savage in his native land is an event which can never be forgotten. Many of my excursions on horseback through wild countries, or in the boats, some of which lasted several weeks, were deeply interesting; their discomfort and some degree of danger were at that time hardly a drawback, and none at all afterwards. I also reflect with high satisfaction on some of my scientific work, such as solving the problem of coral islands, and making out the geological structure of certain islands, for instance, St. Helena. Nor must I pass over the discovery of the singular relations of the animals and plants inhabiting the several islands of the Galapagos Archipelago, and of all of them to the inhabitants of South America.

As far as I can judge of myself, I worked to the utmost during the voyage from the mere pleasure of investigation, and from my strong desire to add a few facts to the great mass of facts in Natural Science. But I was also ambitious to take a fair place among scientific men—whether more ambitious or less so than most of my fellow-workers, I can form no opinion.

Toward the close of our voyage I received a letter whilst at Ascension, in which my sisters told me that Sedgwick had called on my father, and said that I should take a place among the leading scientific men. I could not at the time understand how he could have learned anything of my proceedings, but I heard (I believe afterwards) that Henslow had read some of the letters which I wrote to him before the Philosophical Society of Cambridge, and had printed them for private distribution. My collection of fossil bones, which had been sent to Henslow, also excited considerable attention amongst palaeontologists. After reading this letter, I clambered over the mountains of Ascension with a bounding step and made the volcanic rocks resound under my geological hammer. All this shows how ambitious I was; but I think that I can say with truth that in after years, though I cared in the highest degree for the approbation of such men as Lyell and Hooker, who were my friends, I did not care much about the general public. I do not mean to say that a favorable review or a large sale of my books did not please me greatly, but the pleasure was a fleeting one, and I am sure that I have never turned one inch out of my course to gain fame.

In September, 1858, I set to work by the strong advice of Lyell and Hooker to prepare a volume on the transmutation of species, but was often interrupted by ill health, and short visits to Dr. Lane's delightful hydropathic establishment at Moor Park. I abstracted the MS. begun on a much larger scale in 1856, and completed the volume on the same reduced scale. It cost me thirteen months and ten days' hard labor. It was published under the title of the *Origin of Species,* in November, 1859. Though considerably added to and corrected in the later editions, it has remained substantially the same book.

It is no doubt the chief work of my life. It was from the first highly successful. The first small edition of 1250 copies was sold on the day of publication, and a second edition of 3000 copies soon afterwards. Sixteen thousand copies have now (1876) been sold in England; and considering how stiff a book it is, this is a large scale. It has been translated into almost every European tongue, even into such languages as Spanish, Bohemian, Polish, and Russian. It has also, according to Miss Bird, been translated into Japanese, and is there much studied. Even an essay in Hebrew has appeared on it, showing that the theory is contained in the Old Testament! The reviews were very numerous; for some time I collected all that appeared on the *Origin* and on my related books, and these amount (excluding newspaper reviews) to 265; but after a time I gave up the attempt in despair. Many separate essays and books on the subject have appeared; and in Germany a catalogue or bibliography on "Darwinismus" has appeared every year or two.

The success of the *Origin* may, I think, be attributed in large part to my having long before written two condensed sketches, and to my having finally abstracted a much larger manuscript, which was itself an abstract. By this means I was enabled to select the more striking facts and conclusions. I had, also, during many years, followed a golden rule, namely, that whenever a published fact, a new observation or thought came across me, which was opposed to my general results, to make a memorandum of it without fail and at once; for I had found by experience that such facts and thoughts were far more apt to escape from the memory than favorable ones. Owing to this habit, very few objections were raised against my views which I had not at least noticed and attempted to answer.

It has sometimes been said that the success of the *Origin* proved "that the subject was in the air," or "that men's minds were prepared for it." I do not think that this is strictly true, for I occasionally sounded not a few naturalists, and never happened to come across a single one who seemed to doubt about the permanence of species. Even Lyell and Hooker, though they would listen with interest to me, never seemed to agree. I tried once or twice to explain to able men what I meant by Natural Selection, but signally failed. What I believe was strictly true is that innumerable well-observed facts were stored in the minds of naturalists ready to take their proper places as soon as any theory which would receive them was sufficiently explained. Another element in the success of the book was its moderate size; and this I owe to the appearance of Mr. Wallace's essay; had I published on the scale in which I began to write in 1856, the book would have been four or five times as large as the *Origin,* and very few would have had the patience to read it.

I gained much by my delay in publishing from about 1839, when the theory was clearly conceived, to 1859; and I lost nothing by it, for I cared very little whether men attributed most originality to me or Wallace; and his essay no doubt aided in the reception of the theory. I was forestalled in only one important point, which my vanity has always made me regret, namely, the explanation by means of the glacial period of the presence of the same species of plants and of some few animals on distant mountain summits and in the arctic regions. This view pleased me so much that I wrote it out *in extenso,* and I believe that it was read by Hooker some years before E. Forbes published his celebrated memoir on the subject. In the very few points in which we differed, I still think that I was in the right. I have never, of course, alluded in print to my having independently worked out this view.

Hardly any point gave me so much satisfaction when I was at work on the *Origin* as the explanation of the wide difference in many classes between the embryo and the adult animal, and of the close resemblance of the embryos within the same class. No notice of this point was taken, as far as I remember, in the early reviews of the *Origin,* and I recollect expressing my surprise on this head in a letter to Asa Gray. Within late years several reviewers have given the whole credit to Fritz Müller and Häckel, who undoubtedly have worked it out much more fully, and in some respects more correctly than I did. I had materials for a whole chapter on the subject, and I ought to have made the discussion longer; for it is clear that I failed to impress my readers; and he who succeeds in doing so deserves, in my opinion, all the credit.

This leads me to remark that I have almost always been treated honestly by my reviewers, passing over those without scientific knowledge as not worthy of notice. My views have often been grossly misrepresented, bitterly opposed and ridiculed, but this has been generally done, as I believe, in good faith. On the whole I do not doubt that my works have been over and over again greatly overpraised. I rejoice that I have avoided controversies, and this I owe to Lyell, who many years ago, in reference to my geological works, strongly advised me never to get entangled in a controversy, as it rarely did any good and caused a miserable loss of time and temper.

Whenever I have found out that I have blundered, or that my work has been imperfect, and when I have been contemptuously criticized, and even when I have been overpraised, so that I have felt mortified, it has been my greatest comfort to say hundreds of times to myself that "I have worked as hard and as well as I could, and no man can do more than this." I remember when in Good Success Bay, in Tierra del Fuego, thinking (and, I believe, that I wrote home to the effect) that I could not employ my life better than in adding a little to Natural Science. This I have done to the best of my abilities, and critics may say what they like, but they cannot destroy this conviction.

My *Descent of Man* was published in February, 1871. As soon as I had become, in the year 1837 or 1838, convinced that species were mutable productions, I could not avoid the belief that man must come under the same law. Accordingly I collected notes on the subject for my own satisfaction, and not for a long time with any intention of publishing. Although in the *Origin of Species* the derivation of any particular species is never discussed, yet I thought it best, in order that no honorable man should accuse me of concealing my views, to add that by the work "light would be thrown on the origin of man and his history." It would have been useless, and injurious to the success of the book, to have paraded, without giving any evidence, my conviction with respect to his origin.

But when I found that many naturalists fully accepted the doctrine of the evolution of species, it seemed to me advisable to work up such notes as I possessed, and to publish a special treatise on the origin of man. I was the more glad to do so, as it gave me an opportunity of fully discussing sexual selection—a subject which had always greatly interested me. This subject, and that of the variation of our domestic productions, together with the causes and laws of variation, inheritance, and the intercrossing of plants, are the sole subjects which I have been able to write about in full, so as to use all the materials which I have collected. The *Descent of Man* took me three years to write, but then as usual some of this time was lost by ill health, and some was consumed by preparing new editions and other minor works. A second and largely corrected edition of the *Descent* appeared in 1874.

My book on the *Expression of the Emotions in Men and Animals* was published in the autumn of 1872. I had intended to give only a chapter on the subject in the *Descent of Man*, but as soon as I began to put my notes together, I saw that it would require a separate treatise.

My first child was born on December 27, 1839, and I at once commenced to make notes on the first dawn of the various expressions which he exhibited, for I felt convinced, even at this early period, that the most complex and fine shades of expression must all have had a gradual and natural origin. During the summer of the following year, 1840, I read Sir C. Bell's admirable work on expression, and this greatly increased the interest which I felt in the subject, though I could not at all agree with his belief that various muscles had been specially created for the sake of expression. From this time forward I occasionally attended to the subject, both with respect to man and our domesticated animals. My book sold largely; 5267 copies having been disposed of on the day of publication.

I am not conscious of any change in my mind during the last thirty years, excepting in one point presently to be mentioned; nor, indeed, could any change have been expected unless one of general deterioration. But my father lived to his eighty-third year with his mind as lively as ever it was, and all his faculties undimmed; and I hope that I may die before my mind fails to a sensible extent. I think that I have become a little more skillful in guessing right explanations and in devising experimental tests; but this may probably be the result of mere practice, and of a larger store of knowledge. I have as much difficulty as ever in expressing myself clearly and concisely; and this difficulty has caused me a very great loss of time; but it has had the compensating advantage of forcing me to think long and intently about every sentence, and thus I have been led to see errors in reasoning and in my own observations or those of others.

There seems to be a sort of fatality in my mind leading me to put at first my statement or proposition in a wrong or awkward form. Formerly I used to think about my sentences before writing them down; but for several years I have found that it saves time to scribble in a vile hand whole pages as quickly as I possibly can, contracting half the words; and then correct deliberately. Sentences thus scribbled down are often better ones than I could have written deliberately.

Having said thus much about my manner of writing, I will add that with my large books I spend a good deal of time over the general arrangement of the matter. I first make the rudest outline in two or three pages, and then a larger one in several pages, a few words or one word standing for a whole discussion or series of facts. Each one of these headings is again enlarged and often transferred before I begin to write *in extenso*. As in several of my books facts observed by others have been very extensively used, and as I have always had several quite distinct subjects in hand at the same time, I may mention that I keep from thirty to forty large portfolios, in cabinets with labeled shelves, into which I can at once put a detached reference or memorandum. I have bought many books, and at their ends I make an index of all the facts that concern my work; or, if the book is not my own, write out a separate abstract, and of such abstracts I have a large drawer full. Before beginning on any subject I look to all the short indexes and make a general and classified index, and by taking the one or more proper portfolios I have all the information collected during my life ready for use.

I have said that in one respect my mind has changed during the last twenty or thirty years. Up to the age of thirty, or beyond it, poetry of many kinds, such as the works of Milton, Gray, Byron, Wordsworth, Coleridge, and Shelley, gave me great pleasure, and even as a schoolboy I took intense delight in Shakespeare, especially in the historical plays. I have also said that formerly pictures gave me considerable, and music very great delight. But now for many years I cannot endure to read a line of poetry: I have tried lately to read Shakespeare, and found it so intolerably dull that it nauseated me. I have also almost lost my taste for pictures or music. Music generally sets me thinking too energetically on what I have been at work on, instead of giving me pleasure. I retain some taste for fine scenery, but it does not cause me the exquisite delight which it formerly did. On the other hand, novels, which are works of the imagination, though not of a very high order, have been for years a wonderful relief and pleasure to me, and I often bless all novelists. A surprising number have been read aloud to me, and I like all if moderately good, and if they do not end unhappily—against which a law ought to be passed. A novel, according to my taste, does not come into the first class unless it contains some person whom one can thoroughly love, and if a pretty woman all the better.

This curious and lamentable loss of the higher aesthetic tastes is all the odder, as books on history, biographies, and travels (independently of any scientific facts which they may contain), and essays on all sorts of subjects interest me as much as ever they did. My mind seems to have become a kind of machine for grinding general laws out of large collections of facts, but why this should have caused the atrophy of that part of the brain alone, on which the higher tastes depend, I cannot conceive. A man with a mind more highly organized or better constituted than mine, would not, I suppose, have thus suffered; and if I had to live my life again, I would have made a rule to read some poetry and listen to some music at least once every week; for perhaps the parts of my brain now atrophied would thus have been kept active through use. The loss of these tastes is a loss of happiness, and may possibly be injurious to the intellect, and more probably to the moral character, by enfeebling the emotional part of our nature.

My books have sold largely in England, have been translated into many languages, and passed through several editions in foreign countries. I have heard it said that the success of a work abroad is the best test of its enduring value. I doubt whether this is at all trustworthy; but judged by this standard my name ought to last for a few years. Therefore it may be worthwhile to try to analyze the mental qualities and the conditions on which my success has depended; though I am aware that no man can do this correctly.

I have no great quickness of apprehension or wit which is so remarkable in some clever men, for instance, Huxley. I am therefore a poor critic: a paper or book, when first read, generally excites my admiration, and it is only after considerable reflection that I perceive the weak points. My power to follow a long and purely abstract train of thought is very limited; and therefore I could never have succeeded with metaphysics or mathematics. My memory is extensive, yet hazy: it suffices to make me cautious by vaguely telling me that I have observed or read something opposed to the conclusion which I am drawing, or on the other hand in favor of it; and after a time I can generally recollect where to search for my authority. So poor in one sense is my memory that I have never been able to remember for more than a few days a single date or a line of poetry.

Some of my critics have said, "Oh, he is a good observer, but he has no power of reasoning!" I do not think that this can be true, for the *Origin of Species* is one long argument from the beginning to the end, and it has convinced not a few able men. No one could have written it without having some power of reasoning. I have a fair share of invention, and of common sense or judgment, such as every fairly successful lawyer or doctor must have, but not, I believe, in any higher degree.

On the favorable side of the balance, I think that I am superior to the common run of men in noticing things which easily escape attention, and in observing them carefully. My industry has been nearly as great as it could have been in the observation and collection of facts. What is far more important, my love of natural science has been steady and ardent.

This pure love has, however, been much aided by the ambition to be esteemed by my fellow-naturalists. From my early youth I have had the strongest desire to understand or explain whatever I observed—that is, to group all facts under some general laws. These causes combined have given me the patience to reflect or ponder for any number of years over any unexplained problem. As far as I can judge, I am not apt to follow blindly the lead of other men. I have steadily endeavored to keep my mind free so as to give up any hypothesis, however much beloved (and I cannot resist forming one on every subject), as soon as facts are shown to be opposed to it. Indeed, I have had no choice but to act in this manner, for with the exception of the Coral Reefs, I cannot remember a single first-formed hypothesis which had not after a time to be given up or greatly modified. This has naturally led me to distrust greatly deductive reasoning in the mixed sciences. On the other hand, I am not very skeptical—a frame of mind which I believe to be injurious to the progress of science. A good deal of skepticism in a scientific man is advisable to avoid much loss of time, [but] I have met with not a few men, who, I feel sure, have often thus been deterred from experiment or observations, which would have proved directly or indirectly serviceable.

My habits are methodical, and this has been of not a little use for any particular line of work. Lastly, I have had ample leisure from not having to earn my own bread. Even ill health, though it has annihilated several years of my life, has saved me from the distractions of society and amusement.

Therefore, my success as a man of science, whatever this may have amounted to, has been determined, as far as I can judge, by complex and diversified mental qualities and conditions. Of these, the most important have been—the love of science—unbounded patience in long reflecting over any subject—industry in observing and collecting facts—and a fair share of invention as well as of common sense. With such moderate abilities as I possess, it is truly surprising that I should have influenced to a considerable extent the belief of scientific men on some important points.

THE GREAT WORK

GEOFFREY WEST

THE ANCHOR was weighed, the last delaying ropes cast off, the destination clearly marked upon the map, yet still progress remained of the slowest. It may be unjust to say he pottered, but that is the impression. There were excuses. Anyone who has ever attempted the arrangement of large masses of scattered and diverse material for more than the most superficial and immediate ends will sympathize with his hesitant deliberation, for even in the most minor cases the impulse is always to probe a little further, to test a little more completely, that the conclusion, when at last it is reached, may be just as final and unassailable as possible. Charles had special reasons for obeying such an impulse. He expected contradiction, attack, abuse. His case, he determined, must have no Achilles' heel.

That mid-January, 1855, the whole family moved to a rented house in London—27 York Place, off Baker Street—seeking both change of air and diversion after a spell of children's illnesses. But suddenly the weather turned bitterly cold. Day after day unyielding frost made the pavements iron and the air ice. Even indoors one could not keep warm, and the streets, often, were thick with whirling snow. The children fell ill again, and Charles and Emma so coughed and sneezed and suffered from rheumatism, first one and then the other, that scarcely once were they able to go out together.

All the family were glad to return to Down on February 15, the road cutting through deep snowdrifts massed and moulded by the wind, the countryside stretching away white and dazzling on either side. Snow, feet deep, covered all the garden, frozen so hard that the children, to their delight, could walk upon it without breaking the smooth pure surface. Illness still dogged them; another epidemic appeared, and in March the house was noisy with whooping cough.

It still seemed to Charles "far the greatest fact" that he had "at last quite done with the everlasting barnacles," but by now he was hard at work collecting his species notes and sending out personal appeals for further information, though with the actual writing of any book on the subject comfortably "two or three years" away. He could, at that distance, regard the project with equanimity; any suggestion of its nearer approach disturbed him instantly. In June and July he was still tentative, even to Hooker: "You ask how far I go in attributing organisms to a common descent: I answer I know not." Only a year later did he return to his earlier assurance and confess to Asa Gray, by then a regular correspondent: "As an honest man, I must tell you that I have come to the heterodox conclusion, that there are no such things as independently created species—that species are only strongly defined varieties."

Through all these months and for many more to follow he was constantly busy with multitudinous experiments directed to support or destroy that conclusion. Tests suggested in one field were painstakingly applied in others. "Horrid puzzles" were faced with patient persistence and that supreme honesty which must reject the most helpful evidence till positive proof speaks in its favor. Geographical distribution was at once his inspiration and his bugbear, in the large view both illuminating and illuminated by his theory, in detail presenting the most perplexing problems as to means of dispersal from common centers of development—the only alternative to multiple creation. There was much talk just then of vanished continents once joining lands now widely separated by deep oceans. It would have solved a great group of Charles's difficulties to have accepted the idea, and he "earnestly" wished he might do so, but could not, believing that the existing divisions of land and water were of the very greatest antiquity. Instead, he gave himself up to the most exhaustive and exhausting experiments to discover the possibility of eggs or seeds being carried from land to land either floating in the sea, or upon or in the bodies of migrating birds. Specimens were kept in tanks of seawater at varying temperatures for varying periods, then tested for germination. He supposed seeds swallowed by a fish, the fish by a heron, the heron flying far afield, the seeds being voided and falling upon fertile soil. Sometimes everything went well, and all his geese were swans; sometimes all went ill, and all his swans were geese—but well or ill he patiently went forward, testing every point, determined to accept nothing lightly. He was taking up botanical studies too, largely for amusement but with an eye on variation, and in 1855 he became, at Yarrell's urging, a pigeon fancier, the birds interesting him as a notable case of extraordinary variation from one known form. He wanted, particularly, to discover whether the young of the various breeds differed as much from one another as did their parents. He developed quite an affection for them, and they became, in the following years, as familiar a sight about Down House as half a century before at The Mount. He duly joined the Philoperistera and Columbarian Clubs of fellow-fanciers, a strange set of odd beings with much of the single-minded intensity of racing touts, meeting in "gin palaces" to discuss with dark solemnity and "awful shakes of the head" the finer mysteries of successful breeding. They treated him with marked respect, addressing him formally (or perhaps informally) as "Squire."

The preliminary work still went on in 1856. It seemed indeed that it might well go on forever without Charles ever coming to the actual point of printing "Chapter One" at the head of a sheet of paper and beginning to write. Sometimes he felt that time was slipping away and he growing old too fast, but always some point would arise which obviously—but *obviously*—must be attended to at once. It was Alfred Russel Wallace who, indirectly in 1856 as directly in 1858, was to spur him forward.

Wallace was fourteen years younger than Charles, a man of poor though literate parentage and indifferent schooling, thrust upon the world to make his own living (at land surveying in the company of an elder brother) at fourteen, and only by hard effort keeping alive his ever-growing interest in natural history. A chance meeting with H. W. Bates led to their joint expedition to the remote headwaters of the River Amazon, in prospect a most dubious venture but in achievement, despite setbacks, extremely successful. He spent four years in South America, from 1848 to 1852, then, following eighteen months in England arranging his collection and writing a book on his travels, he set out, early in 1854, upon an eight-years' visit to the Malay Archipelago.

He had, he said, long been "bitten by the passion for species and their description," and as early as 1847, when he read Chambers' *Vestiges,* he inclined to the idea of a progressive development of animals and plants. Now, resting at Sarawak in February, 1855, while Charles stood at the window of 27 York Place, looking out at the driving snow, he wrote an essay "On the Law which has Regulated the Introduction of New Species." It appeared in the *Annals and Magazine of Natural History* in the following September. Lyell read it there, early in 1856, and called Charles's attention to it, urging him to publish at least a sketch of his views, lest, delaying too long, he should be forestalled.

Charles was perturbed. Lyell was pressing. He had begun to doubt whether Charles would ever write his book at all, and may even have wished to force his friend's hand, knowing that once committed he would have to go forward. He asked for something short—he did not mind how short—just to establish priority. Charles "hated the idea of writing for priority," though he would certainly be "vexed if any one were to publish my doctrines before me." He and Lyell had a long talk on the matter in London, early in May; Hooker was appealed to and supported Lyell, and at last Charles rather unwillingly agreed to give "a couple of months" to sketching out a very brief abstract of his views to be issued as "a *very thin* and little volume."

The writing began on May 14, and for a while went steadily ahead. Bunbury saw him on June 20, and there was talk of approaching publication. But he was as uneasy about it as a cat on hot bricks. Repeatedly he lamented—even twice in the same letter—how "dreadfully unphilosophical" it was to publish without full detail, protesting that *truly* he would never have dreamed of it but for Lyell's persistence. He so expected disapproval from his friends that he attributed it even to the faithful Hooker.

At last the old cautious impulse, the desire to delay before publishing, began again to have its way. By July the idea of the brief outline and the material so far written for it were abandoned, and he was deciding that any essay must be as complete as his existing material would allow. He would publish in, perhaps, a year, confining himself to his notes, rejecting new investigations. But even that he could not keep to, and soon he was back at his salt-water tanks and his skeletons of young pigeons and rabbits. He *must* verify his facts. If the task still got "bigger and bigger with each month's work" he couldn't help it. The job had to be done properly, for this was—he was quite firm now—*the* book that he was writing. He would call it *Natural Selection.*

Charles worked on all that winter at Variation. Then health became troublesome again. He visited, in April, [a] water-cure establishment at Moor Park near Farnham in Surrey. The weather, as all that summer, was superb. He liked the place and people, he lazed, read idly, thought of everything but work. Nevertheless it was from here that, on May 1, he wrote to Wallace for the first time, answering a letter and commenting on the paper in the *Annals* as showing their similarity of thought and views. He mentioned his own work now in progress, saying that he had written "many chapters" but did not expect to go to press for two years, and adding that while he could not explain his views on variation in a letter, still he held "a distinct and tangible idea" of how it came about. That was all; as Wallace later pointed out, he gave no clue even remotely indicating Natural Selection.

He was much more specific in another letter, written to Asa Gray four months later, and subsequently important as containing his only explicit statement to this date of the significance of that principle of divergence which had come to seem to him, next to the primary fact of Natural Selection, the keystone of his theory. It was a very typical letter. Gray, reading his views, would "utterly despise" him. Well, he realized the difficulties and even agreed with some of the objections. Lamarckian talk of habit producing delicate adaptation was quite futile. His own view was different and, he thought, more satisfactory. Human selection, he asserted, had been the main agent in forming our domestic species. Suppose, then, a *natural* selection working not on this or that one feature for a few brief years, but upon the entire organization, and unremittingly over millions of generations, only a minority in each generation surviving to propagate its kind, and constantly exposed to changing environment. "Considering the infinitely various ways beings have to obtain food by struggling with other beings, to escape danger at various times of life, to have their eggs or seeds disseminated, etc. etc., I cannot doubt that during millions of generations individuals of a species will be born with some slight variation profitable to some part of its economy; such will have a better chance of surviving, propagating this variation, which again will be slowly increased by the accumulative action of natural selection; and the variety thus formed will either coexist with, or more commonly will exterminate its parent form." That was the main, basic point, but this matter of divergence was important too, for since "the same spot will support more life if occupied by very diverse forms," divergence favors survival and is thus an intrinsic part of the life process. Conversely, the slighter divergences conflict more than wider ones, so tending to mutual extinction and creating the branching tree of life now specifically declared by Charles as by Wallace. He couldn't, he admitted in conclusion, say much about the laws governing variation itself; these were unimportant save as providing the material that selection itself worked upon.

Variation had his attention all the summer. The Indian Mutiny, though its horrors filled the papers, was as remote as the Crimean War a year or two earlier. He was at Moor Park again in June. Henslow came visiting in August. Work went on. That autumn, probably through the winter too, he was considering Hybridism. In December he wrote to Wallace that the book was now half-finished, but publication still—*still!*—two years away. Wallace had asked: would the book discuss Man? Charles replied that he thought to "avoid the whole subject, as so surrounded with prejudices; though I fully admit that it is the highest and most interesting problem for the naturalist." And again he gave his characteristic wriggle away from responsibility: "My work . . . will not fix or settle anything; but I hope it will aid by giving a large collection of facts, with one definite end."

On he plodded, two or three hours a day all he could manage. From Hybridism he turned to Instinct, and thence, apparently, back once more to Geographical Distribution. By April, 1858, he had, he told Gray, completed "eleven long chapters," but had still to deal with palaeontology, classification, embryology, and some other difficult topics, so that he would hardly be ready for the printer before the spring of 1860 at earliest.

At some such pace, ever delaying, always with publication comfortably two years ahead, he might have gone on almost if not quite to the end of his life had not Wallace, quite unwittingly, intervened on June 18, with his essay, written in the previous February, "On the Tendency of Varieties to Depart Indefinitely from the Original Type." It came by the morning's post, an envelope containing no more than a few thin sheets closely penned over and a note expressing the writer's hope that "the idea would be as new" to Charles as to himself and that "it would supply the missing factor to explain the origin of species," and asking Charles, if he thought it worthy, to forward it to Lyell.

Charles read the essay instantly, and wrote at once to Lyell. His letter was odd, and revealing, not least in its tense restraint. He had received "the enclosed," he said, today. It seemed to him "well worth reading." (He must, as he penned the easy phrase, have felt its appalling irony: the essence of all his long labor set forth upon a featherweight of foreign correspondence paper—"well worth reading"!) His feeling broke out irresistibly in the next sentence: "Your words have come true with a vengeance—that I should be forestalled. I never saw a more striking coincidence; if Wallace had my MS. sketch written out in 1842, he could not have made a better short abstract! Even his terms now stand as heads of my chapters." Well, the honorable thing must be done. Wallace didn't actually suggest publication, but Charles would "of course" at once offer to submit the essay to any journal. Again his strong emotion declared itself: "So all my originality, whatever it may amount to, will be smashed, though "(consoling afterthought) "my book, if it will ever have any value, will not be deteriorated; as all the labor consists in the application of the theory." Then, in the last sentence, control once more fully regained: "I hope you will approve of Wallace's sketch, that I may tell him what you say."

It is a letter raising at once the question of Charles's real feeling about priority, about, to put it more widely, public recognition and applause. He commonly and sincerely protested his indifference to, even his distaste for, concern with such matters. All the "wretchedness" of scientific controversy sprang from desire for fame, he wrote in 1848: "The love of truth alone would never make one man attack another bitterly." Yet he could not, for all that, deny altogether the taint in himself, specifically envying what he saw as Hooker's easier freedom. While on the *Beagle* the joy of breaking new trails, of being the very first to scale some hill or scan some view, was very strong in him. He rejoiced in the thought of priority in collecting in some remote area, was dashed when he heard that he had been preceded. In later stay-at-home, sedentary life this feeling quite naturally transferred itself to his mental exploration. It was, in fact, all quite natural. Men do inevitably desire priority. It is a fundamental human instinct.

Wallace's essay put Charles to the acid test. He did care for "the bauble fame," he had confessed in 1857; he wished he could care less for it. Consequently some conflict in him was as inevitable as manifest. It expressed itself in his distracted waverings, his frantic appeals to Lyell and then to Hooker, his conscience-stricken desire, expressed in a letter to Wallace half-written and then destroyed, to suppress his own work completely, leaving the entire field to the newcomer. The depth of the tumult going on in him is clear in a letter to Lyell. Desperately he wanted to accept Lyell's and Hooker's protests against his self-effacement, *if* he could feel he was doing so "honorably." (The word occurs three times in a few sentences.) He put the points for himself: that there was nothing in Wallace's essay not to be found, "much fuller," in his own sketch of 1844, that Asa Gray had his letter, "a year old," showing that he had anticipated *all* Wallace's essential points. He would "now" be "extremely glad" to publish a sketch of his theory, but since he had not done so before, since this eagerness sprang only from Wallace's intervention, weren't, he questioned, his hands tied? Back and fore he swayed between thoughts of such publication and the conviction that even to dream of it was "base and paltry." He concluded in final abasement by apologizing for troubling Lyell with such "a trumpery letter, influenced by trumpery feelings," upon so "trumpery" an affair. He was only sending it, he said, in order to be able to banish the whole subject from his mind for a time—yet next day was sending another note post-haste after it, declaring the same qualms in very much the same language. Desperately he was anxious to do the right thing; desperately he was anxious also that his years of work should find their recognition.

One can scarcely doubt that in any case the only fair solution, some form of simultaneous publication, would have been arrived at. But at this point the whole matter was removed from Charles's hands. Scarlet fever had spread from the village to Down House, and within little more than a few hours the youngest child, his namesake, was dead. Henrietta was sickening, apparently from diphtheria, and very soon two of the attendant nurses were also ill. Charles was distracted, busying himself to get the other children safely out of the house, comforting his wife, reproaching himself more than ever for his "miserable" worry over priority.

Meanwhile Hooker was acting with decision. He had sent to Charles demanding Wallace's essay and the copy of the letter to Gray by instant return, and these were delivered to him by hand at Kew on the evening of Tuesday, June 29, together with the sketch of 1844. Mrs. Hooker at once set to work to transcribe the relevant portions of both letter and sketch, and two days later, on the evening of July 1, these, with the essay, were read to a meeting of the Linnean Society by the Secretary. Hooker and Lyell were present, and spoke shortly, but more to impress the audience with the importance of the matter than anything else.

The extracts from the 1844 sketch were read first, stating briefly but plainly Charles's essential conception of the working of natural and also sexual selection. Then came the passages from the Gray letter, adding the vital principle of divergence, necessary to be stated here since Wallace set it in the forefront of his essay, appearing even in the title, and dominating the argument.

So, that evening of July 1, 1858, the theory of Evolution by Natural Selection was publicly announced.

The audience, Hooker said, was intensely excited, but there was no discussion; at most murmurs and hesitant talk between individual members when all was over. They had come there that evening in anticipation of a paper by Bentham asserting the fixity of species, to hear, in effect, its total contradiction. They might have scoffed, but the plea of Hooker and still more of Lyell had made that difficult; they could only think their thoughts and go their ways, leaving the whole matter to secure burial in the pages of the *Journal of the Proceedings of the Linnean Society.* That in fact seemed likely to happen, for in the scientific world generally the event aroused scarcely more attention than Wallace's other paper in the *Annals* three years earlier. But one straw served to show that a wind *was* blowing—Bentham's immediate withdrawal of his essay, by no means in assent to these new views, but in instant recognition that the subject had been put upon an entirely new footing.

The meeting came out into the London streets under a night sky in which a comet, symbol of fate and change, burned whitely. Those who noted it did so purely astronomically. The days of destinies read in the stars were dead forever now. Yet change, inevitable, irresistable, was at hand. They had just witnessed the release of a pebble which in the next few years would grow to an avalanche.

Charles was glad to hear the proceedings had all gone "prosperously" that evening. Doubts still worried him whether he had acted honorably, and wished that Hooker would "exonerate" him by writing direct to Wallace. But his mind was made up now on one matter. The thing was done.

The Malayan feline had forced his cat out of its bag forever. Delay belonged to the past. He itched to get on with a publishable abstract of that larger work which itself had begun as abstract also.

First, though, he and the family must get away from sickbeds and sickrooms, to recover health and vitality. They all went first to Elizabeth's at Hartfield, then to the Isle of Wight. It was there, at Sandown, on July 20, two days after the death of his sister Marianne, that he received the Linnean Society *Journal* proof sheets of his and Wallace's papers, the reading of which induced him to abandon his holiday intention. He started work on the abstract that very day.

Even then he was doubting whether it could be confined to the usual thirty-page maximum of *Journal* papers, and soon it seemed that the opening section alone would fill as much. He felt that either he would have to compensate the Society for extra printing expenses, or else offer the work as not one but a series of papers.

He was, for once, enjoying the writing, but it irked him to think of having to give his conclusions without the support of all available evidence, and he wondered whether such presentation might not destroy interest in the full work, untouched since the receipt of Wallace's essay but which he meant to take up again at first opportunity.

He wrote daily, at Sandown and then at Shanklin, till August 12, when the return to Down interrupted him for more than a month. But with every day spent upon it the manuscript grew not only in fact but also in prospect. By early October the series of papers had become "a small volume" involving another four (in November five) months' steady labor. By the end of the year he had in view a volume of up to five hundred pages.

By Christmas the first ten chapters, based almost entirely on material already put into shape for the Great Work, were finished, and he ready to embark in two chapters upon the wide topic of Geographical Distribution, writing almost entirely from memory. That was done by the end of February. Between March 2 and 16 he wrote the thirteenth chapter, on "Mutual Affinities of Organic Beings; Morphology; Embryology; Rudimentary Organs," and before the month's end the last of all, the "Recapitulation and Conclusion," was also completed, and he busily correcting the entire manuscript, to have it ready for the printer early in May.

He was tired, but content to have completed a full outline of the work which had engaged him so long, and which he had come (with Lyell) almost to doubt would ever reach an end. Now the whole field was charted, and with that as basis he surely could not fail to go ahead. The abstract and the full exposition in print, he would be able to feel, he thought, his life's work done. He was happier about Wallace too, having had two cordial letters from him approving the Linnean Society proceedings and the idea of the abstract.

His hopes were rising. In December Hooker had suggested a grant to assist publication, but Charles even then felt that it might "sell enough to pay expenses," and by April he was anticipating something like popularity and "a fairly remunerative sale," though he begged Hooker not to repeat his opinion lest the event falsify it and he be made a laughing stock. John Murray next stepped in, on April 1, to give color to his expectation by unseen acceptance of the book on Lyell's recommendation and a list of chapter titles. He even offered the terms—two-thirds of the net profits—he customarily gave to Lyell himself. Charles thought the decision precipitate, but by April 10 Murray had seen the first three chapters and given final confirmation. He questioned only the title, objecting to the word "abstract," which was accordingly deleted. He also queried the phrase "Natural Selection," thinking it unfamiliar and likely to convey little save to the initiate, but Charles could not surrender that! The original title was *An Abstract of an Essay on the Origin of Species and Varieties through Natural Selection.* In the printed volume it became *On the Origin of Species by Means of Natural Selection, or the Preservation of Favored Races in the Struggle for Life.*

With publication settled Charles may have thought himself done with the book, but the book had not done with him. Although Miss G. Tollet, authoress-sister of Emma's closest friend, and "an excellent judge of style," found no fault with it, Hooker and his wife suddenly declared revision necessary to remove obscurities of statement. Charles was horrified, for on his life, he said, "no nigger with lash over him could have worked harder at clearness than I have done." He was exclaiming how he longed to be rid of the whole thing, when, in mid-May, the punctilious Murray appeared with yet more perturbing proposals. He had been reading the complete manuscript, and his conservative Lyellism—he was a keen amateur geologist—found its essential theory "as absurd as though one should contemplate a fruitful union between a poker and a rabbit." In doubt he had consulted the Rev. Whitwell Elwin, editor of the *Quarterly Review,* who shared Charles's qualms about presenting the theory without full evidence, and proposed—Murray passing on the suggestion—that the author should publish instead a full statement of his observations of pigeons accompanied by a quite brief account of his general views and the promise of a forthcoming larger work which would substantiate them for other creatures. Murray liked the idea, for "everybody is interested in pigeons." Naturally Charles would not consider it. He had done his best, and for the present—till the big book was completed—was determined to stand or fall by the existing abstract.

Murray accepted his decision, but the very thought of reshaping his work, joined to the certain need for a running textual revision, came as a last straw, and on May 18 he collapsed and had to go away for water treatment and complete rest. He read *Adam Bede* for relaxation, and was charmed by it. He returned early in June to find the proofs of the book awaiting him, and to plunge into a prolonged and wearisome task of correction which kept him hard at work (but for another brief hydropathic interlude in July) until September. He travailed, he groaned, he apologized. His emendations covered the galley slips and ran over on to attached sheets. The style, he now agreed, was miserable—incredibly, inconceivably bad. He wrote to Lyell and Hocker what a splendid book "a better man" might have made of his material, how wretched his own feeble effort. Thanking Wallace for the opportunity of reading a new essay, he regretted that he could not profit by it; he must not add "one word" to his book—only to try to improve the prose.

By September 30 he had finished.

Of course the book had its reticences, its omissions, its glosses, its limitations. It was written by man, not God. But it is by any standard a mighty product of the human mind. Question its validity, deny its truth, and still it stands a master work, a synthesis of a whole section of knowledge such as only a handful of beings have achieved in the history of the world. Few books have so instantaneously and so impressively made their mark, and few have deserved to. The indicated fact (and fact it was despite Charles's assertion that up to 1859 he could never meet a single naturalist "who seemed to doubt about the permanence of species") that the idea of evolution—even, from Wells and Matthew to Wallace and Herbert Spencer, of Natural Selection—was "in the air," explained something of the impression it made, but by no means all. The most was in the book itself. "Abstract" perhaps, it was the fruit of more than twenty years of close observation and intensive thought, deriving from all the author's activities of that period with extraordinary comprehensiveness. Charles had that genius which is a capacity for taking pains; he had also that which consists in the power to draw, effectively, constructively, upon a vast range of thought and knowledge. All his work in South America, the Falklands, the Galapagos Archipelago, New Zealand, Australia; all his innumerable experiments, from floating seeds in sea water to comparing the skeletons of young animals and birds; all his wide reading from breeders' reports to accounts of old voyages; all his botanical, geological, zoological, palaeontological inquiries, all his speculations on rarity and extinction, areas of subsidence and elevation, processes of fertilization, action of environment, transmission of acquired characters, adaptation, correlation, the range and variety of species—these were but some, a few, of the preoccupations whose results were poured into these packed pages.

The book's power, its sheer compulsion, comes in no small degree from its character as a condensation, a selection of a far larger body of facts and views. The realization that behind each actual instance wait not one but a dozen, a score, perhaps a hundred, more, gives it a dynamic quality as striking as its pure logical exposition, which, for all that, is primary and very forceful, the more so for being logic applied with imagination. The reader is told: "He who will go thus far, if he find on finishing this treatise that large bodies of facts, otherwise inexplicable, can be explained by the theory of descent, ought not to hesitate to go further, and to admit that a structure even as perfect as the eye of an eagle might be formed by natural selection, although in this case he does not know any of the transitional grades. His reason ought to conquer his imagination." But rather, in fact, is the converse true. The reader's imagination must conquer his reason, for—possibly a dangerous admission—the conception must be grasped in its entirety before it can be fully appreciated in its detail. Keen logical argument, massed proof, free imagination—these give the work its power. It can never be, in the common phrase, *dry*, for it is impregnated with a sense of the minute variety of life, and yet, with that variety, of order and unity. Intrinsically, essentially, an assertion, a demonstration, of order amid apparent chaos, of harmony in heterogeneity, and as such, more than as anything else, irresistible in its apparent achievement, it yet never loses the sense of life's incessant, stimulating wonder.

Charles above all felt the book as liberating. Nowhere, save by a slip of the mind or pen, would he commit himself to anything like rigid assertion. Again and again he confessed his ignorance, refusing to fetter himself or any other by what he did not know. Facing the infinitely complex and sensitive balance of nature, he realized the necessity of preserving a basically open mind.

He never pretended to have explained everything, or indeed more than a process: he specified that he was not treating of origins as such, and in that connection came quite quickly to regret his misleading title. He freely admitted that no change of species could be directly proved, that his case rested on what was ultimately circumstantial evidence. He sought only to claim that, if his theory raised difficulties, it solved still more; it afforded a more direct, economical, and thereby harmonious explanation, better because simpler, than any creationist view. It remains that to the present day; more, it remains, despite all, and some very serious, questioning of its adequacy, the most satisfying total account of its broad subject yet put forward. That is why it holds still, against all attacks, its essential place. Even the possibility that its limitations, enforced by vulgar misunderstanding, may make it a very dangerous book, cannot change that fact. We do perhaps most urgently need today another, a deeper-visioning Darwin: he has not yet arisen.

In the following selection we turn to the work of the Austrian monk who was the founder of the science of genetics and who thus ranks only a step below Darwin in importance. It is an irony of scientific history that one of its greatest discoveries, first announced only six years after the publication of The Origin of Species *and offering specific answers to problems neither Darwin nor his immediate successors could solve, should not have been recognized until thirty-five years later, in 1900. The life and work of this unassuming experimenter and its importance for the study of heredity are here discussed by Mendel's fellow-countryman and biographer, who died in 1952.*

GREGOR MENDEL AND HIS WORK

HUGO ILTIS

IT IS 120 years since, in a small village on the northern border of what was called Austria at that time, a boy was born in a farmer's house who was destined to influence human thoughts and science. Germans, Czechs, and Poles had settled side by side in this part of the country, quarreling sometimes, but mixing their blood continually. During the Middle Ages the Mongolic Tatars invaded Europe just there. Thus, the place had been a melting pot of nations and races, and, like America, had brought up finally a splendid alloy. The father's name was Anton Mendel; the boy was christened Johann. He grew up like other farmers' boys; he liked to help his father with his fruit trees and bees and retained from these early experiences his fondness for gardening and bee-keeping until his last years. Since his parents, although not poor compared with the neighbors, had no liquid resources, the young and gifted boy had to fight his way through high school and junior college (Gymnasium). Finally he came to the conclusion, as he wrote in his autobiography, "That it has become impossible for him to continue such strenuous exertions. It was incumbent on him to enter a profession in which he would be spared perpetual anxiety about a means of livelihood. His private circumstances determined his choice of profession." So he entered as a novice the rich and beautiful monastery of the Augustinians of Bruenn in 1843 and assumed the monastic name of Gregor. Here he found the necessary means, leisure, and good company. Here during the period from 1843 to 1865 he grew to become the great investigator whose name is known to every schoolboy today.

On a clear cold evening in February, 1865, several men were walking through the streets of Bruenn toward the modern school, a big building still new. One of these men, stocky and rather corpulent, friendly of countenance, with a high forehead and piercing blue eyes, wearing a tall hat, a long black coat, and trousers tucked in top boots, was carrying a manuscript under his arm. This was Pater Gregor Mendel, a professor at the modern school, and with his friends he was going to a meeting of the Society of Natural Science where he was to read a paper on "Experiments in Plant Hybridization." In the schoolroom, where the meeting was to be held, about forty persons had gathered, many of them able or even outstanding scientists. For about one hour Mendel read from his manuscript an account of the results of his experiments in hybridization of the edible pea, which had occupied him during the preceding eight years.

Mendel's predecessors failed in their experiments on heredity because they directed their attention to the behavior of the type of the species or races as a whole, instead of contenting themselves with one or two clear-cut characters. The new thing about Mendel's method was that he had confined himself to studying the effects of hybridization upon single particular characters, and that he didn't take, as his predecessors had done, only a summary view upon a whole generation of hybrids, but examined each individual plant separately.

The experiments, the laws derived from these experiments, and the splendid explanation given to them by Mendel are today not only the base of the modern science of genetics but belong to the fundamentals of biology taught to millions of students in all parts of the world.

Mendel had been since 1843 of the brethren of the beautiful and wealthy monastery of the Augustinians of Bruenn, at that time in Austria, later in Czechoslovakia. His profession left him sufficient time, and the large garden of the monastery provided space enough, for his plant hybridizations. During the eight years from 1856 to 1864, he observed with a rare patience and perseverance more than 10,000 specimens.

In hybridization the pollen from the male plant is dusted on the pistils of the female plant through which it fertilizes the ovules. Both the pollen and the ovules in the pistils carry hereditary characters which may be alike in the two parents or partly or entirely different. The peas used by Mendel for hybridization differed in the simplest case only by one character or, better still, by a pair of characters; for instance, by the color of the flowers, which was red on one parental plant and white on the other; or by the shape of the seeds, which were smooth in one case and wrinkled in the other; or by the color of the cotyledons, which were yellow in one pea and green in the other, etc. Mendel's experiments show in all cases the result that all individuals of the first generation of hybrids, the F_1 generation as it is called today, are uniform in appearance, and that moreover only one of the two parental characters, the stronger or the dominant one, is shown. That means, for instance, that the red color of the flowers, the smooth shape of the seeds, or the yellow color of the cotyledons is in evidence while the other, or recessive, character seems to have disappeared. From the behavior of the hybrids of the F_1 generation, Mendel derived the first of the experimental laws, the so-called "Law of Uniformity," which is that all individuals of the first hybrid generation are equal or uniform. The special kind of inheritance shown by the prevalence of the dominant characters in the first hybrid generation is called alternative inheritance or the pea type of inheritance. In other instances, however, the hybrids show a mixture of the parental characteristics. Thus, crossing between a red-flowered and a white-flowered four o'clock *(Mirabilis)* gives a pink-flowered F_1 generation. This type of inheritance is called the intermediate, or *Mirabilis,* type of inheritance.

Now, Mendel self-pollinated the hybrids of the first generation, dusting the pistils of the flowers with their own pollen and obtained thus the second, or F_2 generation of hybrids. In this generation the recessive characters, which had seemingly disappeared, but, which were really only covered in the F_1 generation, appeared again and in a characteristic and constant proportion. Among the F_2 hybrids he found three red-flowered plants and one white-flowered plant, or three smooth-seeded and one-wrinkled-seeded plant, or three plants with yellow cotyledons and one with green ones. In general, the hybrids of the F_2 generation showed a ratio of three dominant to one recessive plant. Mendel derived from the behavior of the F_2 generation his second experimental law, the so-called "Law of Segregation." Of course, the characteristic ratio of three dominant to one recessive may be expected only if the numbers of individuals are large, the Mendelian laws being so-called statistical laws or laws valid for large numbers only.

The third important experimental law Mendel discovered by crossing two plants which distinguished themselves not only by one but by two or more pairs of hereditary characters. He crossed, for instance, a pea plant with smooth and yellow seeds with another having green and wrinkled seeds. The first, or F_1, generation of hybrids was of course uniform, showing both smooth and yellow seeds, the dominant characters. F_1 hybrids were then self-pollinated and the second hybrid, or F_2, generation was yielded in large numbers, showing all possible combinations of the parental characters in characteristic ratios and that there were nine smooth yellow to three smooth green or three wrinkled yellow to one wrinkled green. From these so-called polyhybrid crossings, Mendel derived the third and last of his experimental laws, the "Law of Independent Assortment."

These experiments and observations Mendel reviewed in his lecture. Mendel's hearers, who were personally attached to the lecturer as well as appreciating him for his original observations in various fields of natural science, listened with respect but also with astonishment to his account of the invariable numerical ratios among the hybrids, unheard of in those days. Mendel concluded his first lecture and announced a second one at the next month's meeting and promised he would give them the theory he had elaborated in order to explain the behavior of the hybrids.

There was a goodly audience, once more, at the next month's meeting. It must be admitted, however, that the attention of most of the hearers was inclined to wander when the lecturer became engaged in a rather difficult algebraical deduction. And probably not a soul among the audience really understood what Mendel was driving at. His main idea was that the living individual might be regarded as composed of distinct hereditary characters, which are transmitted by distinct invisible hereditary factors—today we call them genes. In the hybrid the different parental genes are combined. But when the sex cells of the hybrids are formed the two parental genes separate again, remaining quite unchanged and pure, each sex cell containing only one of the two genes of one pair. We call this fundamental theoretical law the "Law of the Purity of the Gametes." Through combination of the different kinds of sex cells, which are produced by the hybrid, the law of segregation and the law of independent assortment can be easily explained.

Just as the chemist thinks of the most complicated compound as being built from a relatively small number of invariable atoms, so Mendel regarded the species as a mosaic of genes, the atoms of living organisms. It was no more nor less than an atomistic theory of the organic world which was developed before the astonished audience. The minutes of the meeting inform us that there were neither questions nor discussions. The audience dispersed and ceased to think about the matter—Mendel was disappointed but not discouraged. In all his modesty he knew that by his discoveries a new way into the unknown realm of science had been opened. "My time will come," he said to his friend Niessl.

Mendel's paper was published in the proceedings of the society for 1866. Mendel sent the separate prints to Carl Naegeli in Munich, one of the outstanding biologists of those days, who occupied himself with experiments on plant hybridization. A correspondence developed and letters and views were exchanged between the two men. But even Naegeli didn't appreciate the importance of Mendel's discovery. In not one of his books or papers dealing with heredity did he even mention Mendel's name. So, the man and the work were forgotten.

When Mendel died in 1884, hundreds of mourners, his pupils, who remembered their beloved teacher, and the poor, to whom he had been always kind, attended the funeral. But although hundreds realized that they had lost a good friend, and other hundreds attended the funeral of a high dignitary, not a single one of those present recognized that a great scientist and investigator had passed away.

The story of the rediscovery and the sudden resurrection of Mendel's work is a thrilling one. By a peculiar, but by no means an accidental, coincidence three investigators in three different places in Europe, DeVries in Amsterdam, Correns in Germany, Tschermak in Vienna, came almost at the same time across Mendel's paper and recognized at once its great importance.

Now the time has arrived for understanding, now "his time had come" and to an extent far beyond anything of which Mendel had dreamed. The little essay, published in the great volume of the Bruenn Society, has given stimulus to all branches of biology. The progress of research since the beginning of the century has built for Mendel a monument more durable and more imposing than any monument of marble, because not only has "Mendelism" become the name of a whole vast province of investigation, but all living creatures which follow "Mendelian" laws in the hereditary transmission of their characters are said to "Mendelize."

As illustrations, I will explain the practical consequences of Mendelian research by two examples only. The Swede, Nilsson-Ehle, was one of the first investigators who tried to use Mendelistic methods to improve agricultural plants. In the cold climate of Sweden some wheat varieties, like the English square-hood wheat, were yielding well but were frozen easily. Other varieties, like the Swedish country wheat, were winter-hard but brought only a poor harvest. Nilsson-Ehle knew that in accordance with the Mendelian law of independent assortment, the breeder is able to combine the desired characters of two different parents, like the chemist who combines the atoms to form various molecules or compounds. He crossed the late-ripening, well-yielding, square-hood wheat with the early-ripening, winter-hard, but poor-yielding Swedish country wheat. The resulting F_1 generation, however, was very discouraging. It was uniform, in accordance with Mendel's first law, all individuals being late-ripening and poor-yielding, thus combining the two undesirable dominant characters. In pre-Mendelian times the breeder would have been discouraged and probably would have discontinued his efforts. Not so Nilsson-Ehle, who knew that the F_1 generation is hybrid, showing only the dominant traits, and that the independent assortment of all characters will appear only in the F_2 generation. Self-pollinating the F_1 plants he obtained an F_2 generation showing the ratio of nine late-ripe poor-yielding to three late-ripe well-yielding, to three early-ripe poor-yielding, to one early-ripe, well-yielding wheat plants. The desired combination of the two recessive characters, early-ripe, well-yielding, appeared only in the smallest ratio, one in sixteen—but because recessives are always true-breeding, or as it is called "homozygous," Nilsson-Ehle had only to isolate these plants and to destroy all others in order to obtain a new true breeding early-ripe and well-yielding variety which after a few years gave a crop large enough to be sold. Thus, by the work of the Mendelist, Nilsson-Ehle, culture of wheat was made possible even in the northern parts of Sweden and large amounts heretofore spent for imported wheat could be saved.

Another instance shows the importance of Mendelism for the understanding of human inheritance. Very soon after the rediscovery of Mendel's paper it became evident that the laws found by Mendel with his peas are valid also for animals and for human beings. Of course, the study of the laws of human heredity is limited and rendered more difficult by several obstacles. We can't make experiments with human beings. The laws of Mendel are statistical laws based upon large numbers of offspring, while the number of children in human families is generally small. But in spite of these difficulties it was found very soon that human characters are inherited in the same manner as the characters of the pea. We know, for instance, that the dark color of the iris of the eye is dominant, the light blue color recessive. I remember a tragi-comic accident connected with this fact. At one of my lecture tours in a small town in Czechoslovakia, I spoke about the heredity of eye color in men and concluded that, while two dark-eyed parents may be hybrids in regard to eye color and thus may have children both with dark and blue eyes, the character blue-eyed, being recessive, is always pure. Hence two blue-eyed parents will have only blue-eyed children. A few months later I learned that a divorce had taken place in that small town. I was surprised and resolved to be very careful even with scientifically proved statements in the future.

Even more important is the Mendelian analysis of hereditary diseases. If we learn that the predisposition to a certain disease is inherited through a dominant gene, as diabetes, for instance, then we know that all persons carrying the gene will be sick. In this case all carriers can be easily recognized. In the case of recessive diseases, feeblemindedness,* for instance, we know that the recessive gene may be covered by the dominant gene for health and that the

* Not all feeblemindedness is inherited. Some cases are due to accidents or falls, some to disease.—Ed.

person, seemingly healthy, may carry the disease and transmit it to his children.

With every year the influence of Mendel's modest work became more widespread. The theoretical explanation given by Mendel was based upon the hypothesis of a mechanism for the distribution and combination of the genes. Today we know that exactly such a mechanism, as was seen by the prophetic eye of Mendel, exists in the chromosome apparatus of the nucleus of the cells. The development of research on chromosomes, from the observations of the chromosomes and their distribution by mitosis to the discovery of the reduction of the number of chromosomes in building the sex cells and finally to the audacious attempt to locate the single genes within the chromosomes, is all a story, exciting as a novel and at the same time one of the most grandiose chapters in the history of science. A tiny animal, the fruit fly, *Drosophila,* was found to be the best object for genetical research. The parallelism between the behavior of the chromosomes and the mechanism of Mendelian inheritance was studied by hundreds of scientists, who were trying to determine even the location of the different genes within the different chromosomes and who started to devise so-called chromosome maps.

Correns, Baur, and Goldschmidt in Germany; Bateson and his school in England; Devries in Holland; Nilsson-Ehle in Sweden, are the outstanding geneticists of the first decade after 1900. But soon the picture changed. The Carnegie Institution for Genetic Research in Long Island, under the leadership of Davenport and later under Blakeslee, became one of the world's centers of genetic research. In 1910, T. H. Morgan, then at Columbia University, later at the California Institute of Technology, started his investigations with the fruit fly, *Drosophila,* and founded the largest and most active school of geneticists. The U.S. Department of Agriculture with its network of experimental stations connected with more than a hundred agricultural colleges became the most admirable organization for breeding of better crops and farm animals based upon the principles of Mendelism. The ideas developed by Mendel have found a new home here in the new world.

From 1905 to 1910, I tried by lectures and by articles to renew the memory of Mendel in my home country and to explain the importance of Mendelism to the people. This was not always an easy task. Once I happened to be standing beside two old citizens of Bruenn, who were chatting before a picture of Mendel in a bookseller's window. "Who is that chap, Mendel, they are always talking about now?" asked one of them. "Don't you know?" replied the second, "it's the fellow who left the town of Bruenn an inheritance!" In the brain of the worthy man the term "heredity" had no meaning, but he understood well enough the sense of an inheritance or bequest.

Darwin's theory of evolution, complemented by Mendel's discoveries in genetics, has withstood the most rigorous examination and has become the cornerstone of modern biology. In the following article, the Professor of Animal Genetics at the University of Edinburgh describes with lucidity and charm the fundamental principles of the theory, the evidence in its favor, the ways in which it helps explain the myriad aspects of life around us, and the false doctrine of inheritance of acquired characteristics with which it has been encumbered. So firmly has this concept of evolving, of flux and change, become entrenched in the thinking processes of modern man that it has itself evolved from a theory applicable only to living organisms to one which has influenced anthropology, archaeology, art, astronomy, and so on through the alphabet of human activity.

EVOLUTION

C. H. WADDINGTON

IN BIOLOGY there are two main theories which bring together so many aspects of living processes that they can be considered worthy of the title *fundamental biology*. One of them is concerned with the smallest units into which biological entities can be analyzed. This is the theory of genetics. The other major theory is of a different kind. Whereas in genetical theory we try to analyze living things into ultimate simple constituents, in this other type of theory —which is the theory of evolution—we try to account for the whole complexity of the living world in all its detail and diversity. The genetical theory corresponds, in a way, to the atomic theory and the later theories of fundamental particles in chemistry and physics. Evolutionary theory, on the other hand, corresponds to the astronomical theories which try to account for the whole structure of the universe.

The theory of evolution has only recently been adopted by science. *The Origin of Species,* more than anything else, persuaded the scientific world that the theory of evolution is justified. However, ideas related to evolution have a long history in human thought. There have always been two main ways in which man has seen the animals and plants by which he is surrounded. According to one tradition of thought, all living creatures are of equal importance; the world is conceived of as a static affair, or one which changes, if at all, only in cycles which continually repeat one another through the ages. In the other tradition of thought, living things are considered to be arranged in an order from lower to higher. Simple creatures such as plants or worms or snails are thought of as being in some way inferior to snakes and mice and cats, and these are again "lower" than, say, horses or monkeys or perhaps lions, which were often considered the noblest of the animals. The first tradition is perhaps more characteristic of Asian thought, the second of European. But the distinction is by no means clear-cut, and one can find examples of Asians who thought in the second way and Europeans who thought in the first.

Throughout most of historical times, men who believed that animals are arranged in an order did not think that this indicated any historical process, by which a higher type of animal is derived from a lower and earlier one. During the Middle Ages in Europe, for instance, men spoke of the Great Chain of Being. This was a system in which the animals were arranged in order, from the lower types, such as sponges and jellyfish, through all the intermediates to the highest forms, such as lions, and eventually man. But they were considered all to have been created by God simultaneously.

It was not till the eighteenth century that some biologists began to consider seriously the possibility that the order in which animals are arranged really indicates the way in which they have been derived one from the other. In this notion, the idea of a static order, created once and for all, becomes changed into that of an evolutionary order, in which the higher steps have been derived by the transformation of the lower. Ideas of this kind were discussed by many biologists in the eighteenth century. Perhaps the most famous of these was the French biologist Lamarck. He argued strongly in favor of the theory of evolution, but unfortunately he put forward one argument which is quite definitely incorrect. Because of this mistake, many people rejected his ideas completely, and refused to agree that evolution has really occurred. Before considering further the nature of Lamarck's mistake, it will be well to look briefly at the evidence which led him, and after him Darwin, to believe that the animals and plants which we see around us today have come into being by a process of evolution, rather than by being created in their present form.

There were four main arguments:

(1) The argument from comparative anatomy. In fish, frogs, birds, and mammals the structure of the kidneys is based on the same general pattern, which involves three

different kidneylike organs, the pronephros, the mesonephros, and the metanephros. Different elements of the pattern are developed in the different types of animals. The fact that they all share a common pattern strongly suggests that the animals are related to one another, and it is possible to construct a scheme in which the frogs are derived from the fish and the birds and the mammals are derived from the frogs. A scheme of this kind would be easy to understand if we could assume that there had been a gradual evolution—*i.e.*, a series of successive modifications, from the fish into the other groups, leading to the development first of the mesonephros, and then of the metanephros.

This is only one example of an enormous number of situations in which animals and plants share some common plan, which is differently developed in the different types. Such facts make it very tempting to suppose that one type of being has been derived from another by a change of emphasis in the various elements in their common pattern.

(2) The argument from embryology. This argument is similar to, and reinforces, the argument from comparative anatomy. When we study the development of a pattern of structures which is shared by many organisms, we again find relations which would be very easy to explain if evolution had occurred. "Recapitulation" is really evidence that modifications of the developmental pathways have taken place. Since many of the early stages in the pathway have important parts to play in development, such as induction, modifications will usually take the form of additions to the existing pattern. In fact our first argument (from comparative anatomy) is seen to represent only a time slice of the embryological argument.

(3) The argument from geographical distribution. One of the things which had the strongest influence in persuading Darwin of the truth of evolution was something which he observed in the Galapagos Islands. On these islands Darwin found a large number of species of birds which do not occur anywhere else. They look something like finches, but are really quite different from the finches known in other parts of the world.

Darwin asked himself why a special set of birds should

have been formed on these islands. Moreover, why should there be so many different species, each of which was very well suited to a particular way of life? There was one kind with a large, heavy beak suitable for opening nuts, another with a sharp beak suitable for picking up insects out of crevices, others suited for eating soft fruits, and so on. Darwin felt that it was not reasonable to suppose that a whole set of special birds, each adapted to a special way of life, should have been created for this particular isolated group of islands. It was much more reasonable to suppose that, ages ago, a small number of birds from the mainland of South America had reached the islands, and then in this isolated situation gradually changed, by evolutionary processes, so that they gave rise to types suited to all the possible ways of life which the islands offered.

Another similar case, on a larger scale, is that of the animals in Australia. This region of the world has been cut off by deep seas from any other land for a long period. The animals which are native to Australia belong to a special (and as we now think, evolutionarily primitive) type of mammal called the *marsupials*. They are characterized by bringing forth their young at a very early stage of their development, so that after their birth they have to be carried around for some time by their mother in a special pouch, where they are nourished by milk from her milk glands. Animals of this type are rare in most of the world, but in Australia nearly all the native animals are of this kind. There we find animals which are marsupials, but which otherwise resemble in many ways the different types of animals which are found in the other continents. For instance, there is a marsupial wolf, which lives in a wolflike way, marsupial rats, and marsupial animals that live in trees like squirrels, as well as marsupial types such as the kangaroo, which lives on grass like cattle but has, of course, quite a different structure. Again, the easiest way to explain this is to suppose that some animals of marsupial type once reached the Australian continent, and then, being isolated from the rest of the world, evolved into all the types suitable to live in the circumstances of that continent.

(4) The argument from the fossil record. The strongest

and most direct argument for evolution is to be found in the nature of the remnants which have been preserved in the rocks of the creatures which lived at earlier stages in the earth's history. When sand or mud is washed down by rivers into the sea, it sinks to the bottom as a layer in which shells or bones or other hard parts of the animals may be included. In time these layers condense and harden into rocks. Thus in the rocks we can find the remnants of earlier living creatures, with the earliest at the bottom and more recent ones piled on top. Careful study of these fossils shows us some, at least, of the changes which living things have undergone as the ages have passed. Of course, many of the most important parts of animals and plants are made of soft materials which are not preserved; and again there are many organisms which very rarely, if ever, get buried in the sand or mud of the sea bottom. The record provided by the fossils is therefore an incomplete one. But, in spite of this, it is sufficient to show that evolutionary changes have really happened, and to give us much information about what these changes actually were. For instance, in very early rocks we find remnants of fish but never of amphibia like frogs, or reptiles like snakes or crocodiles, or birds or mammals. These groups make their appearance only in rocks of later ages. Amphibia come next after fish, then reptiles, and after them birds and mammals. This is the same sequence which we had been led to suggest from the studies of comparative anatomy and embryology. So that these three lines of evidence all agree in suggesting the same general course of evolution.

In spite of this evidence, there were two difficulties which originally prevented people from accepting the theory of evolution. One was of a very general kind. Religious thought in Europe had for many centuries considered that the whole world had been created at a single point in time, or at least during a very short period, such as the six days mentioned in the book of Genesis. It was difficult for men to adopt the evolutionary point of view, which regards the world as being continually in a process of creation, which has not ceased yet and which will continue into the future. From the more narrow point of view of science itself, the main difficulty was to discover the causes for the evolutionary changes. This was the difficulty which Darwin solved with his theory of natural selection. This theory is essentially a very simple one, and once it had been stated the argument is so clear and so difficult to deny that it was strong enough to overcome the prejudices of those who wished to continue believing in a single act of creation.

The basis of Darwin's argument was this. He pointed out that the individuals of any population of animals show slight differences from one another. Among horses some will be more fleet of foot than others; among monkeys some will be stronger and more active, and so on. Some of these differences may depend on the particular circumstances in which the animal has grown to maturity. Animals which grow up under favorable conditions of climate and food are likely to be stronger than those which have to struggle against famine and drought. But the external situation does not account for all the differences; some of them are due to inborn differences—*i.e.*, the hereditary potentialities of the different individuals.

Now, Darwin argued that, since the animals of any one kind differ from one another, some of them will be more successful in leaving offspring to make the next generation. Those who do leave offspring in this way will pass on to their progeny those parts of their advantage which are due to their hereditary constitution. It took the insight of a genius to realize this point for the first time. But it is easy now, after we have read Darwin's argument, to acknowledge its truth. So long as individual animals differ from one another in hereditary characters which affect the number of offspring they leave, then the next generation is bound to contain more of the hereditary potentialities derived from the parents who leave most offspring than of those from parents who leave fewer offspring.

It is this process which is known as *natural selection*. This name really compares the processes which go on among wild animals in nature with those which a farmer may carry out with his cattle or sheep or pigs. A farmer will practice "artificial selection." That is to say, he will choose his best bulls and best rams to become fathers of the next generation of his animals. Nature, Darwin urged, does essentially the same thing, because inevitably some animals are more effective than others in transmitting their hereditary potentialities to later generations.

The process of natural selection is often summed up in the phrase "the fittest survive." Now, of course, the word "survive" means really "survive into the next generation by transmitting their hereditary potentialities." Few people make any mistake in understanding this. However, the word "fittest" has led to many misinterpretations. It seems to imply that the animals that are favored by natural selection are those which are strongest, or most fleet of foot, or most powerful in fighting their enemies. It has, in fact, often been interpreted in this way, and the phrase "the survival of the fittest" has been taken to justify the belief that the natural processes of evolution are always on the side of the strong and powerful. But this is quite a mistake. The point that was made by Darwin, and by later biologists, is not really that those who are strong and powerful in their lifetime necessarily have the main influence on evolution. The biological theory is that *those who leave most offspring* are the main influence in determining the character of the next generation.

When a biologist speaks of fitness in relation to evolution he does not mean "fitness" as it is usually understood, in the sense of strength, or success in the general affairs of life. He is referring only to a special sort of success, namely success in leaving offspring. Often, of course, it is those individuals who are fittest, in the general sense of strength and power, who are also fittest in the special biological sense of leaving most progeny, but this need not be so. It may be that animals (or men?) who are very successful in their lifetime are less effective than their humbler colleagues in producing the offspring who will build up the next generation. When this is so it will be, as was pointed out in the Sermon on the Mount, the humble and meek who inherit the earth. One must never forget that biological and evolutionary fitness is not the same thing as worldly success in general.

By drawing attention to the process of natural selection, Darwin pointed out how populations of animals are bound to change their hereditary properties as time passes, and thus are bound to evolve. Like all scientific theories, however, this one solved certain problems only to direct attention to other questions, which required a further answer. One of the most important questions that Darwin's theory raised was this. How does it come about that all the individuals of the same species are slightly different from one another in their hereditary capacities? Darwin himself had no very clear answer to this. He pointed out that we can observe that different individuals do differ, but he was not certain how these differences had arisen. In his time the theory of genetics and of biological heredity had not yet been formulated.

Darwin sometimes fell back on the explanation which had been suggested earlier by the French biologist Lamarck. According to him, the hereditary differences between individuals arise initially from the different conditions in which they grow and develop. It is a well-known observation that, if an organism has to carry out its life under certain particular difficulties, it frequently becomes modified in such a way to make it easier to meet these stresses. For instance, a plant which is grown in poor soil and with little water will grow slowly, into a stunted form, and produce its seeds for the next generation when it still has only few leaves and a short stem. It becomes modified, in fact, so that it carries out its essential evolutionary function of producing progeny in a way which makes as much use as possible of the slender resources available to it. Again, a horse which is continually practiced in racing acquires the ability to run faster. Or, to take a human case, a man who carries out heavy muscular work becomes stronger and better able to do it.

Lamarck supposed that these modifications, acquired during the lifetime of an individual, in response to the particular stresses it has to meet, are passed on to its offspring. This theory is known as *the inheritance of acquired characters*. Like Darwin's theory of natural selection, it is a very simple idea, and many people are at first sight tempted to believe it. However, it is really quite unlike Darwin's theory of natural selection, since there is no logical necessity about it. We cannot say from general principles that characters acquired during the lifetime *must* be hereditary, and passed on to the next generation. In fact, the more one thinks about it, the more difficult it is to think of any possible mechanism by which they could be passed on in this way. A blacksmith uses his arms a lot and acquires strong muscles; how can this affect the sperm cell which he contributes to the next generation? Darwin himself was very doubtful if it would do so, although at one point in his career he invented a special theory about how it might happen. However, the question is one which can be solved only by experiments which give us some understanding of the general processes of heredity and of the way in which external stresses may or may not affect them.

In the period since Darwin propounded the theory of natural selection we have, of course, developed a large body of knowledge about heredity. This is the science of genetics. We know that hereditary potentialities are carried from one generation to the next in the form of the genes in the chromosomes of the gametes. We know also that these genes have a strong tendency to preserve their nature unchanged, whatever the character of the body of the animal in which they are carried. We know very many examples in which a modification of the body is made artificially and is then found not to be transmitted to the next generation. In many breeds of dogs it is the common practice to cut off the tip of the tail in the puppies, and in some groups of men, such as Jews, the practice of circumcision has been carried out for generations. In neither case does the modification of the individual have any effect on his progeny. Changes in heredity only occur when the genes themselves are changed.

The mere "acquiring of a character" by an individual during its life is rarely enough to bring about changes in the genes. In recent years a school of Russian geneticists have stated that, under special circumstances, such alterations may sometimes occur. They have stated that if "winter wheat," which is normally sown in the autumn so as to ripen in the following summer, is instead sown in the spring, it will become converted into a heritable variety which ripens in the summer immediately after sowing, without having to go through the period of cold weather in winter. Most scientists, are still not convinced that such changes really happen. They point out that in the Russian experiments it was not quite certain that the winter wheat which was sown did not include some grains of wheat of a spring variety. Certainly when attempts have been made in other countries to repeat the Russian experiments these have nearly always been unsuccessful. It is, of course, very difficult to prove experimentally that such effects *cannot* occur. It might be that they happen only very rarely, under very particular conditions.

Moreover, one can imagine some processes which might produce effects which would easily be mistaken for hereditary changes, although not really being so. The reproduction of a plant such as wheat involves the formation of a seed, which contains a great deal of nutritive material as well as the actual germ cell from which the new individual will arise. It might be possible, by treating a plant, to influence the kind of nutritive material which it forms in its seeds, and this might influence the way the germ cell will develop when the seed grows. In this way, a treatment of the parent would influence the character of its offspring, without actually changing its hereditary potentialities. Some biologists in other countries believe that the Russians may have come across effects of this kind; very few of them are ready to accept the Russian claim that they have altered the hereditary potentialities by changing external conditions, even in rare and peculiar situations.

Certainly it is very generally believed, amongst biologists in all countries, that as a rule the modifications which upbringing may produce in an animal or plant during its lifetime have no effect on the potentialities it transmits to the next generation. Lamarck's mistake was to think that they usually do have an effect. It is very fortunate that Lamarck was wrong. If he were right, and if heredity were commonly altered by conditions of life, then we should find that those animals and men who were unfortunate enough to live under difficult and unfavorable circumstances would have had their hereditary potentialities damaged. As a result, their offspring would show the effects of the bad conditions to which their parents had been subjected. Fortunately we do not have to fear this. However bad the conditions of life may be for an individual, his offspring will, at their conception, have full and undamaged hereditary potentialities.

Although we can be very confident that the hereditary materials are very rarely, if ever, altered by the conditions of life of the animal that bears them, there is still very much that we do not understand about the processes of gene change, or mutation. We know that such changes take place spontaneously: that is to say, they occur in animals and plants to which nothing special has been done, but which are living their normal healthy lives. Under such circumstances, any particular gene mutates very seldom, perhaps once in 100,000 or 1,000,000 generations. In unicellular organisms, one generation is equal to the division time of the cell. In multicellular organisms it must, however, be remembered that many cellular divisions may take place between fertilization and the emergence of the gamete cells which begin the next generation. However, each individual contains many genes, certainly some tens of thousands and possibly hundreds of thousands. This means that a change in some one or other of these numerous genes is not so rare, but is going on all the time.

Any population of animals or plants, at the present day, will contain many different hereditary potentialities, which have been produced by mutations in their ancestors. It is these mutations which cause the hereditary variations on which natural selection can operate. If a gene changes, and produces a form which has a harmful effect (that is, an effect which causes the animal showing it to leave fewer offspring), then of course this gene will be passed on rather rarely to later generations, and may soon disappear again. However, new examples of this harmful gene will continually be produced by mutation. In fact, after a time, a balance will be struck between the process of mutation, which causes the appearance of a number of new harmful genes in every generation, and the process of natural selection, by which these genes pass on to later generations only in low numbers and therefore tend to disappear. On the other hand, of course, some gene mutations will produce effects which lead the individual to leave more offspring. These will be favored by natural selection, and will become commoner in later generations.

We might perhaps have expected, or hoped, that Nature would operate in such a way that only favorable gene mutations would be produced. However, this is not what happens. It seems that the mutation, or changing of a gene, is always, in a way, a mistake. Every gene should normally produce another gene exactly like itself whenever the cell divides, and it should remain stable and constant when the cell is merely growing and not dividing. Gene mutations are best regarded as the result of something going wrong, either with the process of identical duplication of the genes, or with their stability.

The gene is an extremely elaborate structure of chemical atoms combined in a very definite pattern. It must contain at least some hundreds of thousands of atoms, all in their particular places. It is not surprising that occasionally something goes wrong, and an alteration occurs in such an elaborate structure. We can, in fact, increase the frequency of such mistakes by acting on the genes with penetrating radiations such as X-rays, which travel into the cell and as it were shake up the genes in its nucleus. A certain number of special chemicals, such as the mustard gases used in warfare, will also affect genes in this way. None of the agents we know of, however, will produce changes of any definite type in the genes; they merely make it more likely that some haphazard alteration will happen, without specifying what kind of alteration this will be or which of the many genes will be affected.

This is usually expressed by saying that gene mutations are random events. This means that we cannot tell exactly when a change in a gene is going to happen, nor what sort of change it will be. In some ways gene mutation may be compared to a game of chance, such as rolling dice. Suppose we are playing with 5 dice whose faces are marked with numbers from 1 to 6, and that we are trying to get a high score. After a time we may have made a throw which has given us a score of, say, 25. If this is the highest we have scored, this will be the arrangement that natural selection would have perpetuated. To do any better than that, we have to break up that arrangement and roll the dice again. This corresponds to a gene mutation. It is easy to see that the chances are that our next throw would be less than 25; that is to say, a gene mutation with harmful effects will have occurred. But, if we are ever to do better than 25, we shall have to put up with a lot of unsuccessful throws while waiting to get one which gives a higher score than we started with. This is what happens in evolution. In all animals and plants, each gene changes occasionally, and the great majority of these changes are harmful. This damage is the price that has to be paid in order that sometimes a useful new gene may be produced.

The differences between individuals of the same species is not always due to the occurrence of new gene mutations. Very often these differences arise because of different combinations of genes which had mutated many generations earlier. A simple illustration may make this clearer. Let us use A B C D, etc., for a number of genes, and the corresponding smaller letters a b c d, etc., for alternative forms into which they may change by mutation. Some generations after these mutations have occurred we may find individuals in which the genes are combined in all sorts of different ways. We may have one diploid animal with Aa Bb CC dd, and another with aa Bb Cc dd, and so on. The biological fitness—*i.e.*, number of offspring they leave—of all these individuals is likely to be different. Each individual is, as it were, an experiment carried out by Nature to try to produce the most fit type. Again, there is a price to be paid if the reshuffling is to go on, so that more and more fit types eventually appear. This price is that individuals should die. If every individual was immortal, there would soon be no room to try out any new types, and improvement would have to stop. The death of individuals, which makes room for new types to appear, and to be tried out, is the only way to avoid stagnation, and is an essential part of the whole process of evolution.

Modern theories, therefore, look on evolution as being very largely the result of chance. New genes, which give rise to new hereditary potentialities, arise by random alternations in the complex substances of the hereditary material; these genes become combined with one another in random and ever-changing ways, to produce the individual animals or plants that we see. There is, at first sight, very little orderliness in the process. Many people find this surprising. But, if we look at the results which evolution has brought about, we shall see that there is not very much orderliness in them either. If you examine a large number of different types of flies, or of beetles, or of many other groups of animals, you will find that there is an extraordinarily large number of varieties, which differ from each other only in minor points, which do not seem to have much importance in the general life of the animal. One species of fly will have six bristles on its back, where another has only four, and so on. These are just the sort of haphazard and senseless-looking alterations which we might expect a random process to produce.

This is, however, not quite the whole story. Although the results of evolution, when looked at in detail, give a general impression of disorderliness and chaos, there are some signs of order also to be discerned in it, and these are particularly interesting. We shall have space here to mention only two aspects of order which may be discerned in evolution. They are both connected with the fact that evolution is brought about by natural selection, that is to say, the ability of organisms to leave offspring. Naturally animals and plants can do this only if they are competent in carrying out the tasks which life demands of them.

Let us look first at an example of evolution, to see what kind of order may be found in it. A suitable example is provided by the evolution of the horse family. The remote ancestors of our modern horses were creatures of quite a different appearance. They were relatively small animals, with feet which possessed three or four toes, and with fairly small teeth which they probably used to eat a diet of the leaves of trees. The bones of such creatures are found in rocks which were laid down some sixty million years ago. In their descendants, whose fossil remains are found in younger rocks, we can see that several types of change took place during the course of evolution. The legs became longer, and the number of toes was reduced to only one, which provides the hoof in modern horses. Again, the general size of the animal increased. Further, the size of the teeth, particularly the crushing molar teeth at the back of the jaw, has become much larger.

Now, all these changes did not go on in all the descendants of the original horses. At any period in the history of the family, we find some kinds which have elongated legs, large teeth, and considerable body size, but other kinds may persist with the old characters, and in others one of the types of change, such as the lengthening of the leg, may have occurred without being accompanied by the others. The three types of change we have mentioned are, in fact, only a general over-all pattern, which one can see when one looks in a broad way at the history of the family as a whole, but which is overlaid by a whole mass of minor changes if one starts to look at the history in detail.

In every species of organism the individuals are, as it were, competing with one another for the production of offspring. Those who leave more offspring contribute more to the next generation, and so the character of the species gradually changes. But each species has also got to carry out its life in a world populated by many other types of plants and animals. If the evolution of a family is to continue through a long period of time, it must find some way of living which is compatible with the various other types of living things which it will meet. The whole family of horses, for instance, evolved into creatures which lived by eating grass and escaping from their enemies by fleetness of foot. But they succeeded in developing this way of life, and becoming efficient at it, only through a long process of random trial and error. Their small three-toed ancestors had lived by nibbling the leaves of trees and shrubs; the later types of horse which persisted in this sort of behavior became eliminated, presumably because they could not do it well enough in the face of the other organisms (such as marauding wolves or tigers, or stronger competitors who wished to eat the same leaves), who made up the whole ecological community of which they were a part.

We can see a reflection of the randomness of gene mutation and recombination in the numerous different modifications of the horse family when it is examined in detail. But, in contrast to this, the general over-all pattern of change, toward a long-legged, big-toothed, fast-running animal, is a reflection of the overriding factor which tends toward orderliness in evolution, namely the necessity for animals and plants to carry out some way of life with maximum efficiency. Organisms, if they are to persist through a long evolutionary history, must become *adapted;* they must, in other words, adopt some definite way of life and become extremely good at doing it. This requirement means that, in spite of the haphazard way in which new hereditary variations appear, the over-all course of evolution will show some orderliness because all the mutations which reduce efficiency, either in the short run for the next generation, or even over moderately long periods of time by leading the family to specialize in some way of life at which another type of animal is more efficient, will eventually be eliminated.

This is perhaps the most important conclusion about evolution. Although it is based on random and disorderly processes, such as gene mutation and the recombination of genes during reproduction, the necessity for organisms to grow up efficiently enough to leave offspring means that in the long run the results of evolution will be to produce an increase in efficiency and effectiveness. This is usually spoken of as *evolutionary progress.*

If we examine one particular group of animals, such as horses, we can see evolutionary progress in respect of their particular mode of life. If we look at the animal kingdom as a whole, with all the numerous different ways of life that different animals follow, can we there also see some general over-all pattern of change which can be considered as progress? Not all biologists, when faced with the problem stated in these definite terms, would agree on the answer. In practice, however, when they are not asked definitely whether or not progress has occurred, they nearly all employ language and concepts which imply that it has done so. The various groups of the animal kingdom are almost always arranged in an order, and the groups in this order are commonly spoken of as the "lower" or "higher." Sponges, worms, and mollusks, for instance, are lower groups, while the vertebrates are of a higher category, and within the vertebrates it is usual to speak of the fish as lower vertebrates, and birds and mammals as higher groups.

It is difficult, however, to express in a few words the general nature of the pattern of evolutionary change which applies to the animal kingdom as a whole. Sir Julian Huxley has spoken of it as an increase in the animal's independence of its environment and in its capacity to live in many different environments. Another attempt to describe the general course of evolutionary change has been to say that it has been an increase in the capacity of animals to react to relations between the things which surround them, rather than to the things themselves. None of these attempts at definition are perhaps very satisfactory. The important point to realize is that evolution, by means of natural selection, will, in the long run, produce a gradual increase in the efficiency with which organisms can find some way of leaving offspring in the face of ever-increasing competition from all the other creatures which are trying to do just the same thing.

In the period since 1900 when Mendelism was introduced to scientists, the vocabulary of genetics has proliferated. To Mendel's relatively simple laws of Uniformity, Segregation, and Independent Assortment have been added the discovery of the chromosomes and genes, the study of radiation and other causes of mutation, and most recently the research on DNA and RNA. In the following selection George W. Beadle first discusses these developments in modern genetics and then surveys the entire field of genetics and its application to organic evolution. His article presents a reasoned survey of past and present, and the facts he adduces offer implications for the future of biology which stagger the imagination. "There is no reason why we cannot . . . direct our own evolutionary futures," says Dr. Beadle. But he warns that "whether we should do this and, if so, how, are not questions science alone can answer. They are for society as a whole to think about."

George Wells Beadle was born in 1903 on a farm near Wahoo, Nebraska. He studied at the University of Nebraska School of Agriculture, became a Ph.D. at Cornell University, and has taught at the California Institute of Technology and Stanford University. He is now President of the University of Chicago. Dr. Beadle received the Nobel Prize in 1958, and has been called "the man who did most to put modern genetics on its chemical basis."

THE PLACE OF GENETICS IN MODERN BIOLOGY

GEORGE W. BEADLE

TWENTY-FIVE years ago Dr. William Morton Wheeler, a distinguished and admired professor of biology and Dean of the Bussey Institution of Harvard University, wrote a small essay in which he said, "Natural history constitutes the perennial rootstock or stolon of biologic science. . . . From time to time the stolon has produced special disciplines which have grown into great, flourishing complexes. . . . More recently another dear little bud, genetics, has come off, so promising, so self-conscious, but alas, so constricted at the base." I am sure Professor Wheeler was convinced that this bud would be abortive.

A few weeks ago there appeared in *Science* a related essay by a distinguished and likewise much admired biologist, Sewall Wright, who was a graduate student at the Bussey Institution during Wheeler's time. After quoting the above words, Wright points out that, far from aborting, the little bud genetics has flourished mightily and has in many respects replaced natural history in the sense that it was become the rootstock of all biological science and has bound "the whole field of biology into a unified discipline that may yet rival the physical sciences."

Why such a change in twenty-six years? For despite the fact that Wheeler was not above giving his friends and colleagues in genetics a bit of ragging, he was basically serious. There has been a great change. We have come to recognize that genetics does in fact deal with the very essence of life.

I should like to begin a development of this thesis that genetics is the keystone of modern biology by reminding you that every one of us—you and I—starts development as a tiny sphere of protoplasm, the almost microscopic fertilized egg; and that somehow in this small sphere there must be contained the specifications, the directions, or the architectural blueprints for making one of us out of that bit of jellylike material. Of course, the process by which this happens is enormously complex, and we do not yet understand very many of the details. But we do know that a substantial part of these directions are wrapped up in the centrally located nucleus of the cell. These directions are the material heredity that we received from our parents.

In addition to this set of directions in the nucleus, there must be more. There must be an architectural organization of the rest of the cell—the cytoplasm—and this is indispensable. And for the carrying out of the directions there must be a proper supply of raw materials in the form of food—perhaps ten or twenty tons—for the egg to grow and differentiate into a mature person. Time, too, is essential—sixteen, twenty, twenty-five years, or more. Finally, there must be a proper environment, initially a very precise one. Later, as we develop the ability to regulate our own environments, we become less fussy. The environment adds to the information in the original egg. This is particularly impressive in our own species, for in addition to all the other environmental information fed into us during development we are continually bombarded with a cultural inheritance—language, art, music, religion, history, science, and so on

—that in man supplements biological inheritance to a far greater degree than in any other species.

All of these factors are essential to our development, and many of them continue throughout life.

Five Questions about Genetic Specifications

What I wish to talk about are the directions in the nucleus. What are they and how do they specify that from this minute cell one of us will come? I shall ask five questions about these specifications:

First, how do we get them and how do we transmit them? I shall dispose of this one briefly, for it is answered by classical genetics—the Mendelian genetics now found in every elementary textbook of modern biology. You know about classical genetics: about blue eyes, brown eyes; curly hair, straight hair; good hemoglobin, bad hemoglobin; and so on.

Perhaps you know less about the remaining four questions:

How are the specifications written—that is, what is the language of genetics?

How are the specifications replicated? From the time we start development as a fertilized egg until we transmit them to the next generation there are perhaps sixteen to twenty-five successive replications of these specifications, depending on whether the carrier is female or male. Each time the material is replicated it doubles, so twenty replications represent more than a million copies. How does replication occur with the precision necessary to avoid intolerable numbers of mistakes?

How are the specifications—the directions or the recipe for making us—translated? This is an enormously difficult question, and I shall say right now that we know very little about it.

How are specifications modified during the course of evolution? Most of us believe in organic evolution, and we want to know how we have come to be different from our ancestors. In other words, what is the nature of the mutation process?

A few years ago, seven or eight years ago, we would have had a very difficult time answering the four questions that I have just asked. We did not know enough, and we did not have many good clues even as to how we might go about searching for answers to these questions. But within the past half-dozen years or so excellent clues have turned up. In 1953 there occurred an important turning point in modern biology. What was it and what does it have to do with answering the questions I have posed?

The Role of Deoxyribonucleic Acid

By this time it had become quite clear to a number of biologists that a particular chemical substance called deoxyribonucleic acid was important in transmitting hereditary information in bacteria and in viruses. Since the cells of all higher plants and animals contain deoxyribonucleic acid, it seemed probable that this substance served to carry genetic specifications in all living systems.

I shall attempt to explain how and why this substance, DNA for short, is important. And I shall try to do it without considering the details of its rather complex chemistry. DNA has been known for a long time. And it was known to consist of long chainlike molecules made of four kinds of units called nucleotides. But it was not known exactly how DNA molecules were internally organized until 1953, when two investigators—Dr. James D. Watson, now at Harvard University, and Dr. Francis H. C. Crick of Cambridge University—succeeded in formulating a structure that has proved to be substantially correct.

From the information then available from classical organic chemistry, from X-ray diffraction studies, from analyses of the relative proportions of the four kinds of nucleotides, and through ingenious model building, Watson and Crick proposed the structure illustrated in Figure IV-1.

This Watson-Crick structure was at once exciting to the biologists. Why? Because it suggested such plausible answers to the four questions: How is genetic information written? How is it replicated? How is it translated? And how does it mutate?

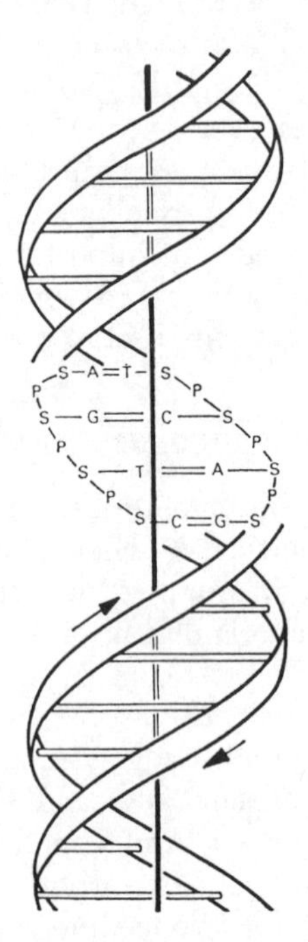

IV-1. *The Watson-Crick structure of DNA schematically represented. The parallel spiral ribbons represent the paired polynucleotide chains. Hydrogen bonding is represented by transverse parallel lines. P = phosphate group, S = sugar unit, A = adenine base, T = thymine base, G = guanine base, C = cytosine base. Arrows indicate that polynucleotide chains run in opposite directions as specified by the sugar-phosphate linkages. Redrawn from Watson and Crick (1953).*

How does the model help answer these questions? The key to the structure of DNA is that its molecules are double in a special way. There are two parallel polynucleotide chains wound around a common axis and bound together through specific hydrogen bonding.

You can more easily visualize the essential features of DNA if you will imagine a four-unit segment of it pulled out in two dimensions as shown in Figure IV-2.

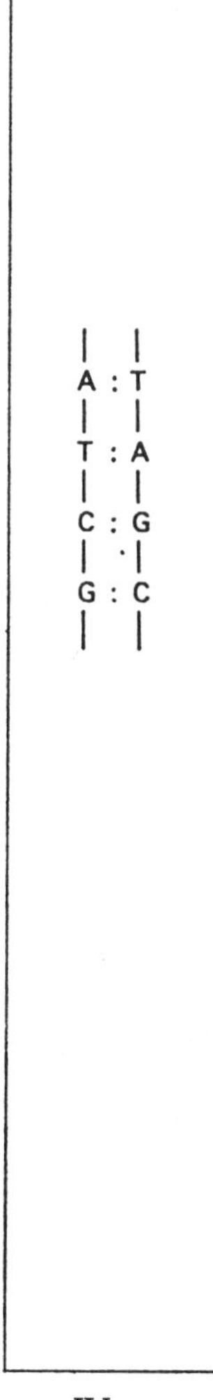

IV-2.

Here the four letters represent the four nucleotides; and the colons, hydrogen bonds. In fact, you can very nicely represent such a segment with your two hands. Place your forearms vertically before you and parallel. Fold your thumbs against your palms and place homologous fingertips together as though they were teeth on two combs vertically oriented in a single plane, tooth tip to tooth tip. In this arrangement the two index fingers represent the A:T nucleotide pair, and so on.

Imagine many fingers along your forearms—of four kinds corresponding to the nucleotides A, T, C, and G. The four kinds of fingers or nucleotides can be arranged in any order on one arm but must always have the complementary order on the other. T opposite A, A opposite T, G opposite C, C opposite G. Thus if one knows the sequence of nucleotides in one chain, the sequence in the other can be determined by the simple rule of complementarity.

This structure suggests that genetic information is contained in the sequence of nucleotides; in other words, DNA is a kind of molecular code written in four symbols. One can think of the code as a sequence of nucleotide pairs or of nucleotides in a single chain, for it is obvious that the double chain and the two single component chains all contain equivalent information. In essence the two complementary chains are analogous to forms of a single message, one written in conventional Morse code, the other in a complementary code in which each dot is changed to a dash and vice versa.

Let us now ask the question: how much information is packed away in the nucleus of a human egg? It is estimated that there are about five billion nucleotide pairs per single cell. How much information does this correspond to in terms of, say, information spelled out in the English language? Francis Crick has expressed it this way: If you were to make an efficient code for encoding messages in English in the four symbols of DNA, and with this started encoding the *Encyclopaedia Britannica* in this DNA code, you could get the contents of the entire set in about two per cent of the amount of DNA in the nucleus of a single fertilized egg cell. Thus, using all the DNA of the egg, you could encode about fifty times that much information. This is another way of saying that it requires the equivalent of about 1000 large volumes of directions in the egg nucleus to specify that a human being like one of us will develop properly from it, given a cytoplasm, proper food, and a suitable environment. Said in another way, that is the size of a genetic recipe for building a person.

This is supposedly the way the genetic information is carried from generation to generation—in a language we might call DNA-ese. Each gene is a segment of DNA of perhaps three or four thousand nucleotides.

Replication Through Double Structure

Now let us ask about the replication. The double structure of DNA suggested immediately to Watson and Crick how this could happen. If, during cell division, the two chains were to come apart, obviously each could serve as a template for picking up additional units to make new half chains. And this is happening in each of us right now. In

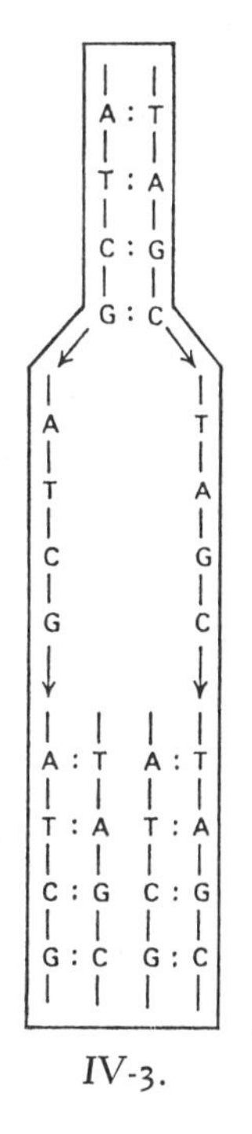

IV-3.

many cells nucleotides are continually being made from food components. The replication of DNA according to this scheme can be illustrated as shown in Figure IV-3.

You can represent the process with your hands. Indicate the double molecule as already directed as paired hands. Take the two hands apart. Imagine free fingers (nucleotides) moving around at random. Each single hand serves to select in proper order the one-fingered units necessary to make a complementary hand. The right hand is a template for making a left hand and vice versa. So with a double molecule, represented by a pair of hands, two single molecules arise by breakage of hydrogen bonds, with each then directing the synthesis of a new complementary single partner.

This process of replication takes place with every cell division and, as we shall see, with a high degree of precision.

This hypothesis by which two identical bipartite molecules arise from a single such double molecule is very satisfying in its simplicity and elegance. If true, it is presumably the basis of all biological reproduction at a molecular level. Can the hypothesis be tested? The answer is yes. In fact, several kinds of experiments can and have been made to see if the hypothesis agrees with observed facts.

In one kind of experiment DNA units are labeled with radioactive phosphorus. Each nucleotide has one phosphorus atom, and a certain number of its phosphorus atoms can be made radioactive by growing an organism, say a bacterium, in a medium containing radioactive phosphorus for several generations until it becomes equilibrated. Then both chains of its DNA molecules will be labeled. If the bacteria are then allowed to multiply in a medium in which there is no radioactivity, the two chains of each DNA molecule, both labeled, should come apart, each then directing the synthesis of an unlabeled partner. The new double molecules should then be labeled in one chain but not in the other. In the next generation the labeled chain should separate from the nonlabeled one. With synthesis of nonlabeled partners by these, there should be produced labeled and nonlabeled double molecules in equal numbers. The observed results are consistent with this expectation.

Another way of doing essentially the same experiment is to replace the normal nitrogen atoms of DNA with "heavy" nitrogen, the stable isotope N^{15} instead of the usual N^{14} counterpart. DNA molecules so labeled become heavier but not larger. Hence they are denser. DNA containing only N^{15} can be cleanly separated from that containing N^{14} in an analytical centrifuge cell in which an appropriate density gradient is established. In such experiments it is found that bacteria containing DNA fully labeled with N^{15} if allowed to multiply once (double in number) in a medium containing only N^{14}, give rise to descendants in which all the DNA molecules are "hybrid" as though one nucleotide chain of the double molecules contained N^{15} and the other N^{14}. This, of course, is what is predicted by the hypothesis. In a subsequent generation, also in N^{14} medium, half the DNA molecules are hybrid and half are fully light. Again, this is what would be expected if the hypothesis is correct.

While experiments of this kind do not prove that the Watson-Crick hypothesis of DNA replication is correct, they do strongly suggest it.

An even more dramatic way of testing the hypothesis is the one used by Professor Arthur Kornberg and his associates, now at Stanford University. They have devised a test-tube system in which there are present the four nucleotides A, T, C, and G as triphosphates, a buffer solution, magnesium ions, and a polymerizing enzyme. DNA molecules added to this system appear to be replicated. Is the new DNA like the primer molecules added? One important observation suggests it is. The ratio of A:T nucleotide pairs to C:G pairs of the product is like that of the primer DNA. It is not easy to see how this could be if the primer were not being copied in a precise way. On the other hand, if DNA having known biological activity (as determined by ability to transform the genetic constitution of a bacterium) is used as a primer, both the product and the primer added end up being inactive. Why this is so is not known, but it is strongly suspected that the polymerizing enzyme added contains a small amount of depolymerizing enzyme that breaks up DNA chains and thus destroys activity.

Again, the Kornberg synthesis does not prove that the

hypothesis is correct. It is just possible that an unkind nature could have evolved a system that would do just exactly what the hypothesis predicts but by a different mechanism.

Ribonucleic Acid as Messenger

About the next question: How is genetic information translated? How do we develop from that minute egg cell? These are enormously difficult questions, and we know relatively little in detail about the answers. They involve the whole of development, differentiation, and function. There are working hypotheses—widely used and useful ones—that suggest how some of the steps occur.

We know that in our bodies there are many thousands of kinds of protein molecules—large, long molecules made of amino acids and very specific in their properties. One, for example, is hemoglobin. It is built of 600 amino acids strung together in a particular way. There are two kinds of chains of amino acids per hemoglobin molecule, each in pairs, with each chain about 150 amino acids long. And we know that there are segments of DNA—two, we postulate—in our chromosomes that say how to build the two protein subunits.

A widely used working hypothesis assumes that around a double helix of DNA there is wound a helix of another kind of nucleic acid, called ribonucleic acid or RNA. RNA, like DNA, is built of four nucleotides. Somehow the DNA code is translated into a corresponding sequence of RNA. RNA then moves from the nucleus into the cytoplasm. There it is incorporated into microsomes, submicroscopic structures in which protein synthesis occurs. In the microsome, RNA units are believed to serve as templates against which amino acids are lined up in proper sequence.

Amino acids, derived from the proteins in our food, are first activated by enzymes and subsequently hooked to small carrier segments of RNA that serve to carry the amino acids to their proper places on the microsomal RNA templates.

Carrier RNA may be thought of as a messenger carrying packages and addresses to which they are to be delivered. The messenger carries the amino acid packages along the

RNA template until the address matches that on the template. There is a specific RNA messenger for each of the twenty kinds of amino acids. When all component amino acids are correctly ordered, they are linked together to form proteins, which then peel off the templates, and the process is set to be repeated. For hemoglobin, for example, there are assumed to be two DNA segments, one for each kind of protein chain, and two corresponding RNA templates. This in essence is believed to be the translation process.

A large number of proteins serve as enzymes or essential components of enzymes. Enzymes catalyze chemical reactions that would otherwise occur at rates so low that life processes would essentially cease. For each enzyme protein there is supposedly a segment of DNA information in the nucleus—a gene—and corresponding microsomal RNA templates in the cytoplasms of those cells active in synthesis of that particular enzyme protein.

An important question of present-day biology is concerned with the nature of the mechanism by which the four-symbol code of DNA is related to the twenty-symbol code of proteins. It is obvious that single symbols of DNA cannot stand for amino acid, for there are only four. Likewise pairs of DNA symbols will not do, for there are only 16 such pairs if the DNA molecule is read in one direction. If one reads in one direction and uses three symbols per amino acid, there are 64 possibilities. However, only 20 of the triplets are useful if successive sets of three are used, for the overlapping sets of three must not encode amino acids or there would be confusion in the translation. Twenty is the minimum number required to encode all of the amino acids that occur in proteins—that is, if one reads the code in one direction. However, because the two parallel chains in a DNA molecule have opposite polarities as determined by the way the nucleotides are oriented in the two chains, the double DNA molecule is symmetrical and there is therefore no obvious way to know in which direction the information is to be read. Unless there exists some kind of marker, as yet undiscovered, that specifies in which direction to read, the number of three-symbol sets that can be used to encode amino acids unidirectionally is only 10. Four-symbol codes have accordingly been investigated. It turns out that there are 27 such four-symbol "words" that can be used without any of their overlaps making sense when read either forward or backward and without the four-letter words themselves making sense when read in reverse. This is sufficient, but it is not known if this is indeed the correct coding mechanism. In fact, there are some indications that it may not be.

I hope this will give you some idea of how investigators go about studying the translation mechanism. A number of workers are busy on it at the present time and many are optimistic that the DNA-protein code will be broken.

Mutations as a Source of Evolution

My fourth question concerns the nature of mutation. How is genetic information modified during the replication in a manner that permits organic evolution?

During DNA replication mistakes are occasionally made. Presumably, during replication a nucleotide does not pick up a complementary partner as it should but instead picks up a noncomplementary one. It has been postulated that such mistakes result from an improbable tautomeric form in which a hydrogen atom is in an improbable position at the exact moment the nucleotide picks up a partner. A wrong partner is therefore selected. In the next round of replication the "wrong" partner will pick up what is its complementary partner, and this will result in substitution of one nucleotide pair for another. This is somewhat like a typographical error. In typographical errors it is possible to have extra letters, too few letters, one letter substituted for another, or transposed letters. Presumably similar kinds of mistakes can be made in genetic information during replication. In fact, there is genetic evidence that these four basic types of mistakes do occasionally occur.

How often do such mistakes occur? Quite infrequently, we believe. From the time one receives a set of directions in the fertilized egg until one transmits it to the next generation —and remember this is perhaps seventeen to twenty successive replications of information equivalent to about 1000 printed volumes—a significant and detectable mistake is made perhaps about once in a hundred times. This is clearly a high order of precision.

What happens to such typographical errors as are made? First of all, it is clear that the DNA molecule will replicate just as faithfully whether the information in it makes sense or not. Its replication is a purely mechanical one, it seems. Therefore mistakes in genetic information will be perpetuated.

It is obvious that if there were no way of eliminating errors in such a process, such errors would accumulate from generation to generation. Perhaps an analogy will make this clear. If a typist types in a purely mechnical way, never proofreading, never correcting, and types successive copies of the same material always from the most recently typed copy, she will accumulate mistakes at a rate dependent on her accuracy until eventually the sense of the original message will be entirely gone. In the same way this would have to happen with genetic information if there were no way of taking care of mistakes. With genetic information something does happen that takes care of mistakes. By extending the analogy perhaps I can make clear what does happen. The typist, typing mechanically, can correct a mistake by a second random typographical error, but obviously the probability of this is extremely low. It is likewise so with genetic information, and it is clear therefore that this is not the principal way in which mistakes are prevented from accumulating. Let us pretend the typist has an inspector standing beside her. When she makes a mistake, he says. "Throw that one away. Put it in the wastebasket and start over." If in the next try she makes no mistake, he says, "All right, now you may type another from the one you have just finished." Each time she makes a perfect copy he allows her to go ahead, but each time she makes a mistake he insists she throw the copy away. That is what happens with genetic information. The inspector is analogous to natural selection. Bad sets of specifications in man are eliminated by natural selection.

A more dramatic term for elimination of unfavorable specifications by natural selection is "genetic death," as used by Dr. H. J. Muller. Individuals developed from unfavorable specifications do not reproduce at the normal rate, and ultimately a line so handicapped dies out. To avoid progressive accumulation of mistakes from generation to generation, it is obvious that every error in replication that is unfavorable must be compensated for by the equivalent of a genetic death. That is why geneticists are concerned about factors that increase the mutation rates.

You may quite properly ask, "Are there no favorable mutations?" The answer is yes, there are occasional favorable mutations; they are, in fact, the basis of organic evolution.

However, because many mutations involve subtle changes that may be favorable under special circumstances of environment or over-all genetic constitution, it is not easy to estimate the proportion of favorable to unfavorable mutations. Theoretical considerations and a certain amount of experimental evidence agree in indicating that the great majority are unfavorable. Organisms are in general already so highly selected for success in their normal environments that the chance of further improvement by random mutation must be very small. Perhaps an analogy with a fine watch will dramatize the point. Assume the watch is very slightly out of adjustment. A random change brought about, say by dropping it, could conceivably improve the adjustment. Clearly, however, the chance of making it run less well or not at all is enormously greater. Now let us extend our typing analogy. Assume our inspector exercises judgment. When the typist makes an error that improves the original message, he passes it. Thus improved messages will replace their ancestral forms and the improvement will be cumulative. Something like this happens with living systems. Specifications improved by occasional favorable mutations are preferentially reproduced and thus tend to replace their ancestral forms. This is natural selection.

In recent years many factors have been found to increase the frequency of mutations. High energy radiation that penetrates to the cell nucleus is mutagenic in proportion to its amount. A number of chemical agents are likewise mutagenic. It is now possible, for example, to alter nucleotides in known chemical ways that will produce mutations. Oxidation of amino groups of nucleotides with nitrous acid is one way. It is encouraging that biochemists and geneticists who study the mechanisms involved are beginning to be able to predict successfully the types of mutations that are most likely to be produced by specific chemical agents. It is not, however, possible to do this specifically for certain genes only.

Let us now turn to the general question of evolution. What do mutations have to do with the processes by which evolution occurs?

Organic evolution is interesting and important in many respects. For one thing, it is not logically possible to accept only a small amount of it, for one cannot imagine a living system that could not have evolved from a very slightly simpler system. Starting with man, for example, and working backward toward simpler systems one sees no obvious stopping place. Our ancestors were presumably a bit simpler than we. Early in man's evolution there were primitive men. And before primitive man there were prehuman ancestral forms capable of evolving into true man. This is true however one defines man. The point is that no matter what living system one thinks of, another is conceivable that is one mutation simpler or different. And so one can go backward in the evolutionary process to simpler and simpler forms until finally one begins to think of systems like present-day viruses, the simplest of which consist of little more than nucleic acid cores (DNA or RNA) and protein coats. One can easily imagine that before systems of this type there were smaller and smaller systems of nucleic acid and protein capable of replication and of mutation which in turn had ancestors consisting of only nucleic acid.

We know that nucleic acids can be built up from nucleotides and these from simpler precursors. In a lecture, Professor Melvin Calvin talked about the origin of some nucleotide precursors and presented evidence suggesting that some such compounds, or their relatives, are found in certain meteorites. It is assumed that these were formed by natural chemical reactions that went on and are still going on outside living systems. Presumably precursors of nucleotides were formed through such reactions. Professor Calvin also mentioned the evidence that amino acids are made from such simple inorganic molecules as methane, ammonia, hydrogen, and water under conditions assumed to have obtained on primitive earth. It is, I believe, justifiable to make the generalization that anything an organic chemist can synthesize can be made without him. All he does is increase the probability that given reactions will "go." So it is quite reasonable to assume that given sufficient time and proper conditions, nucleotides, amino acids, proteins, and nucleic acids will arise by reactions that, though less probable, are as inevitable as those by which the organic chemist fulfills his predictions. So why not self-duplicating viruslike systems capable of further evolution?

I should point out that nucleic acid protected with a protein coat has an enormous selective advantage, for it is much more resistant to destruction than is "raw" nucleic acid. Viruses can be stored for years as inert chemicals without losing the capacity to reproduce when placed in a proper environment. Of course present-day viruses demand living host cells for multiplication, but presumably the first primitive life forms inhabited environments replete with spontaneously formed building blocks from which they could build replicas.

Before molecules like methane, hydrogen, water, and ammonia there were even simpler molecules. Before that there were elements, all of which nuclear physicists and astrophysicists believe have evolved and are now evolving from simple hydrogen. That is why I say if you believe in evolution at all there is no logical stopping place short of hydrogen. At that stage I'm afraid logic, too, runs out.

The story can, of course, be repeated in reverse. When the conditions become right, hydrogen *must* give rise to other elements. Hydrogen fuses to form helium, helium nuclei combine to give beryllium-8, beryllium-8 captures helium nuclei to form carbon, and carbon is converted to oxygen by a similar process. In this and other known ways all the elements are formed. As one goes up the scale, the number of possibilities rapidly increases. As elements begin to interact to give inorganic molecules, the number of possibilities becomes still greater. I do not know how many inorganic molecules are possible, but I do know there must be a very large number. With organic molecules the number becomes truly enormous, particularly with large molecules like proteins and nucleic acids. For example, there are something like 4 raised to the 10,000th power ways a modest sized DNA molecule can be made. There appears to be no stage at which there is a true qualitative change in the nature of evolution. The number of possibilities goes up gradually, the complexity goes up gradually, and there appears to be no point at which the next stage cannot be reached by simple mutation.

Let us suppose we have a small piece of DNA protected by a protein coat and capable of replication in the presence of the proper building blocks and a suitable environment. During replication, the system will occasionally make mistakes. It is a mutable system. Given sufficient time there will eventually occur a combination of mutations of such a nature that the protein coat will become enzymatically active and capable of catalyzing the formation of a nucleotide or amino acid from a slightly simpler precursor. If this particular building block happens to be limiting in replication, the mutant type will obviously have a selective advantage. It can replicate in the absence of an essential building block by making it from a simpler precursor. If two such units with protein coats having different catalytic functions combine to form a two-unit system, they will be able to make two building blocks from simpler compounds and will be able to survive under conditions in which their ancestral forms would fail. In the same way it is not too difficult to imagine systems arising with successively three, four, five, and more units with every additional unit serving a catalytic function. With each additional unit the total system would become one step less dependent on spontaneously performed precursors. With perhaps ten thousand such units the system might be able to build all its necessary parts from inorganic materials as we know present day green plants do.

How many units to reach the stage of man? Perhaps one hundred thousand units carrying out one hundred thousand functions are necessary. However many it is, we know they carry the specifications for the development of a complex nervous system by which we supplement blind biological inheritance with cultural inheritance. We reason, we communicate, we accumulate knowledge, and we transmit it to future generations. No other species we know of does this to anything like the same degree. We have even learned about organic evolution and are on the verge of learning how to start the process.

I pointed out that in the Kornberg system with the four nucleotides present, nothing happens unless a primer is added. That is not entirely true. After a delay of some three or four hours something does happen even without a primer. What happens is that a DNA molecule is spontaneously formed. It differs from all naturally occurring DNA in that it contains only two of the four nucleotides. Now if this two-unit co-polymer is used as a primer in a new system, it immediately initiates the synthesis of co-polymers like itself. In other words, it starts replicating. Remember, it arose spontaneously. If you believe in mutation—and you must if you accept scientific evidence—you must believe that if you start with a two-unit co-polymer and let it undergo successive replications, there will eventually occur a mutation with which a pair of nucleotides will be replaced by the pair originally excluded in the process. This conceivably could have been the origin of the four-unit DNA of all higher organisms.

Knowing what we now know about living systems—how they replicate and how they mutate—we are beginning to know how to control their evolutionary futures. To a considerable extent we now do that with the plants we cultivate and the animals we domesticate. This is, in fact, a standard application of genetics today. We could even go further, for there is no reason why we cannot in the same way direct our own evolutionary futures. I wish to emphasize, however—and emphatically—that *whether* we should do this and, if so, *how*, are not questions science alone can answer. They are for society as a whole to think about. Scientists can say what is possible and perhaps something about what the consequences might be, but they are not justified in going further except as responsible members of society.

The Problem of Ultimate Creation

Some of you will, I am sure, rebel against the kind of evolution I've been talking about. You will not like to believe that it all happened "by chance." I wish to repeat that in one sense it is not chance. As I have said, the mutations by which we believe organic evolution to have occurred are no more "chance" reactions than those that occur in the organic chemist's test tube. He puts certain reactants in with the knowledge that an expected reaction will go on. From the beginning of the universe this has been true. In the early stages of organic evolution the probabilities were likely very small in terms of time intervals we are accustomed to think about. But for the time then available, they were almost certainly not small. Quite the contrary; the probability of evolving some living system was likely high. That evolution would go in a particular direction is a very different matter. Thus the *a priori* probability of evolving man must have been extremely small—for there were an almost infinite number of other possibilities. Even the probability of an organism evolving with a nervous system like ours, was, I think, extremely small because of the enormous numbers of alternatives. I am therefore not at all hopeful that we will ever establish communication with living beings on other planets, even though there may well be many such on many planets. But I do not say we should not try—just in case I am wrong!

Some of you will no doubt be bothered by such a "materialistic" concept of evolution. Ninety years ago in Edinburgh, Thomas Henry Huxley faced this question of materialism in his famous lecture on the physical basis of life. And it has been faced many times since—for example, a few years ago by Dean George Harrison of M.I.T. in his book *What Man May Be*. What Huxley said can be said today with equal appropriateness. He said in effect that just because science must by its very nature use the terminology of materialism, scientists need not necessarily be materialists. A priest wears material clothes, eats material food, and takes his text from a material book. This does not make him a materialist. And so it need not with a scientist. To illustrate, the concept I have attempted to present of the origin of life and of subsequent evolution has nothing to do in principle with the problem of ultimate creation. We have only shifted the problem from the creation of man, as man, to the creation of a universe of hydrogen capable of evolving into man. We have not changed the problem in any fundamental way. And we are no closer to—or further from—solving it than we ever were.

V. The Ways of Living Things

V. *The Ways of Living Things*

During the early days of biology, its practitioners were interested primarily in what is referred to as natural history, the study of plants and animals in their natural habitats. They needed no special equipment, and the nearest field or garden offered multitudes of opportunities. Their principal qualification was the ability to make detailed and accurate observations. Darwin himself, little more than an amateur of natural history before the voyage of the Beagle, *commented on his own abilities: "I think I am superior to the common run of men in noticing things which easily escape attention, and in observing them carefully." The study of natural history is not to be disparaged. From it came the factual information on which the science of biology was founded. In itself, it has resulted in such volumes as Gilbert White's* Natural History of Selborne *and the works of Henri Fabre. And the modern science of ecology is its direct descendant.*

But observation in itself is not enough. The naturalist is struck immediately with the fantastic abundance and diversity of living things. He can quickly become submerged in an ocean of unmanageable facts. Organization and classification thus become essential steps in this as in other sciences. A former member of the faculties of the University of Illinois and of the University of Durham, England, who is now Professor of Geology at University College, Swansea, introduces us to the system of classification devised by the Swedish naturalist Carolus Linnaeus in the eighteenth century, which, with modifications, has continued in use to the present day.

LIVING THINGS

F. H. T. RHODES

The Nature of Life

"WHAT IS LIFE?"—it is a question that all of us ask sooner or later, a question as old as man himself. It is also a question to which there are many answers, and one that is ultimately basic to the whole of human experience. But there is a peculiar sense in which the question is important in a review of the history of living things. What is this common property of life which they all share?

Even this limited form of the question is surprisingly difficult to answer, and the difficulty seems to arise partly because life is unique and cannot therefore be easily defined by analogy or contrast, and partly because it is too complex to be defined in simple concise terms (the same is true of energy and matter). At present we can only describe life as a series of processes which take place within certain complex levels of organization of matter. The most characteristic of these processes are familiar to us all: growth, movement, reproduction, metabolism, irritability, and so on.

"Life" could conceivably exist as a vast number of different processes, each quite unlike that by which it is in fact maintained. But life as we know it is a unity: for all their diversity, countless millions of living things over thousands of millions of years have shared a common life process, and in this lies at once both the simplicity and the wonder of life.

The Abundance and Diversity of Life

THE ABUNDANCE OF INDIVIDUALS. No one who is familiar with living things can fail to be impressed by both their abundance and their variety. The earth, the sky, and the seas literally teem with life, and this abundance is not confined to the more favorable regions such as tropical forests or shallow seas, for even the more inhospitable regions of the earth support rich and varied communities. The abundance of many individual organisms is almost beyond imagination. A recent investigation of the upper one-inch layer of a soil near Washington, D.C., show it to contain more than 1,000,000 macroscopic animals and 2,000,000 macroscopic seeds per acre. A meadow soil of the same latitude proved to contain more than 13,000,000 animals and nearly 34,000,000 seeds per acre. In a California estuary, macroscopic organisms present in the uppermost eighteen inches of sediment numbered more than 3,000,000 individuals per acre. None of these observations took account of the multitude of microscopic forms which were present. Bacteria, for example, abound in many media; a gram of soil may contain several hundred millions.

The prodigality of nature is strikingly shown in the number of eggs produced by many creatures. A single salmon produces as many as 28,000,000 eggs in a season, and a single oyster may "lay" as many as 100,000,000 in a single season. In neither case, however, do more than a very small fraction of these survive; indeed, if all the eggs of the oyster were to be fertilized and developed, and the offspring multiplied under the same conditions, the great-great-grandchildren would number 66,000,000,000,000,000,000,000,000,000,000,000, and the shells of a generation would make a mountain eight times the size of the earth! The oyster is not, however, by any means unique in this respect.

Such is the abundance of life: an abundance which is a major factor in the evolution of living things, because, as Darwin showed, it leads inevitably to "natural selection."

THE DIVERSITY OF LIFE. The abundance of living things is reflected not only in the enormous numbers of individuals which exist, but also in the diversity represented by the great number of species which have been described. More than 300,000 species of living plants are known (about sixty times as many as at the time of Linnaeus) and about 4,750 new plant species are described each year. Of the total number of species, angiosperms (flowering plants) comprise about 150,000 species, thallophytes (algae, fungae, etc.) about 107,000 species, bryophites (liverworts, mosses, etc.) about 23,000 species, and pteridophytes (ferns, horsetails, etc.) about 10,000 species.

The number of known species of living animals is much greater than that of plants. A recent estimate by Mayr puts the number at more than 1,120,000, and, if subspecies are included, the number of named forms is more than 2,000,000. New animal species are being described at a rate of about 10,000 a year. When we think of "animals" we almost instinctively think first of the mammals, the familiar group of generally rather large and conspicuous animals to which we ourselves belong. Only in a rather secondary manner do we consider fish, frogs, lizards, snakes, birds, and so on to be animals. All these groups collectively constitute the vertebrates and, in spite of the fact that to us they "are animals," the vertebrates as a whole constitute only about five per cent of all known animal species. More than three-quarters of all described animal species are insects, while mollusks (snails, mussels, etc.), other arthropods (crabs, spiders, etc.), chordates (vertebrates, etc.), and protozoans (unicellular organisms) (in decreasing order of numerical importance) constitute the other more common groups. Sponges, coelenterates (corals, etc.), the "worm-like" phyla, bryozoans ("moss animals"), and echinoderms (starfish, sea urchins, etc.) are less diverse (coelenterates include about 9,000 recognized species, for example), and the remaining phyla are conspicuously smaller. These relationships are illustrated in Figure V-1.

Even this staggering diversity does not indicate the full variety of living things, however, for within a species there may still be conspicuous variation. Our own species affords a ready, though not an exceptionally varied, example. The nature and extent of this subspecific variation is one of the prime concerns of contemporary taxonomy (the science of classification of organisms), for it appears to be a factor of great importance in the development of new species.

All our discussion has so far been concerned only with living organisms. Life has, however, existed on the earth for

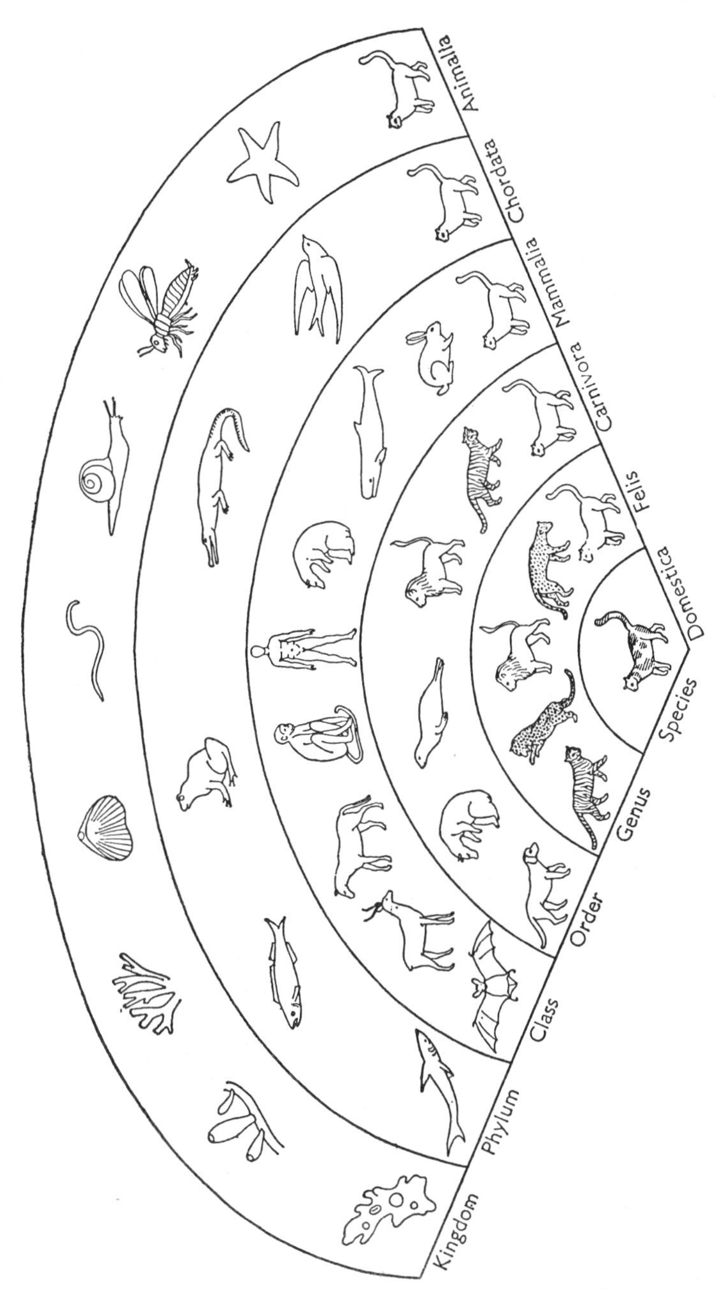

V-1. *Diagram showing the basis of the Linnean system of classification, and illustrating the increasing similarity of members of the various taxonomic categories from the Kingdom to the Species.*

at least 2,700,000,000 years, and probably very much longer. Simpson has recently suggested that the total number of species which may have existed since the "dawn of life" is probably of the order of 500,000,000 and this does not seem to be an excessive estimate. On this calculation more than ninety-nine per cent of all the species that once existed have now become extinct.

The Patterns of Life

INTRODUCTION—SIMILARITIES AND DIFFERENCES. Whenever we wish to refer to events or objects or ideas, we must use names, and this applies to animals and plants no less than to other things. We read that Adam, the first taxonomist, coined names for the animal world and these "kinds" of animals are recognized both by the trained taxonomist and by the casual observer. Mayr records the striking fact that the "primitive Papuan of the mountains of New Guinea recognizes as species exactly the same natural units that are called species by the museum ornithologist."

By the use of such names any form of organism can be designated. We often find it useful, however, to recognize similarities between different kinds of animals or plants. Thus early man doubtless classified the animals around him as "dangerous or harmless" and the plants as "useful or useless" or "edible or poisonous." We wander in a garden and naturally refer to weeds, vegetables, fruit trees, flowers, moss, and so on. Aristotle indicated how animals could be characterized according to their habitat, structure, and habits. Later students classified organisms upon their methods of locomotion (e.g. "creeping things"—worms, mollusks, insects, snakes; "flying things"—insects, birds, mammals, fishes), their feeding habits (e.g. carnivorous or herbivorous), their reproductive characteristics, their body form, their physiological characteristics, and a host of other features.

Each of these different methods of classification is valid and, within certain obvious limits, useful, and the choice of a particular method or basis of classification will depend largely upon the purpose which it is designed to serve. If the classification is to be of value in the detailed study of living and fossil organisms it must fulfill three requirements. Firstly it must provide a distinctive name for each recognizable "kind" of organism and the full name must be such as to indicate both the immediate relationships of the organism and also its specific identity. Secondly, such a classification must be readily applicable and must be useful and intelligible to students throughout the world. Local or dialect names are therefore unsuitable, and a standardized system of nomenclature must be developed and enforced by an accepted code. Thirdly, such a classification must indicate genetic relationships. It must therefore combine individual "kinds" of organisms which are similar, and serve to distinguish them from others from which they differ. To some extent, it must also indicate degrees of relationship. For example, a lion and a tiger are more closely related to each other than either is to a horse, to which, however, they are more closely related than they are to a lobster. These relative differences must be reflected in a satisfactory classification.

THE MODERN BIOLOGICAL METHOD OF CLASSIFICATION. For almost two thousand years after the death of Aristotle (384–322 B.C.) a system of classification based largely upon his concepts was employed by students of living things. The first trend towards the establishment of a more adequate classification was marked by the studies of the English naturalist John Ray (1627–1705) and culminated in the work of the great Swedish botanist Carl von Linné (Carolus Linnaeus 1707–78), whose *Systema Naturae* laid the foundation of our present method of classification. He divided living things into kinds or "species" on the basis of structural characters, and gave to each species a distinctive name. Linnaeus also proposed a hierarchy of higher categories: genus, order, and class.

The organic world may be divided into two broad groups: the animal kingdom, and the plant kingdom. (Some specialists employ a third kingdom, the Protista, to include the problematical unicellular organisms.) Now the conventional division into two kingdoms clearly includes a very large number of greatly diverse organisms, which resemble one another only in a limited number of characters. Each of the kingdoms may, therefore, be further subdivided into successively smaller groups, each of which will share more characters in common, the ultimate limit being reached with the individual organism.

There are seven main categories of classification and these, together with a number of intermediate forms, constitute the taxonomic hierarchy. The successive restriction of each category may be seen in the illustrated examples (Figure V-1), in which the groups are arranged in decreasing order of size. Some indication is also given to the characters which are used to define groups at various taxonomic levels.

The basic taxonomic unit is the species. Neontologists (students of living organisms) differ widely in their definitions of this category but the following definition is commonly accepted: "Species are groups of actually (or potentially) interbreeding populations which are reproductively isolated from other such groups." The problem of recognizing such groups in fossil specimens is complicated both by the incompleteness of the fossil record and by the time factor, which is represented in any sequence of fossils. Some palaeontologists prefer, therefore, to define a species as "an ancestral-descendant sequence of interbreeding populations evolving independently of others, with its own separate and unitary evolutionary role and tendencies." A group of closely related species comprises a genus, a group of genera constitutes a family, a group of families an order, a group of orders a class, and a group of classes a phylum.

Linnaeus also devised the system whereby each species is known by a name which includes two words, the first representing the genus and the second the species. These names are sometimes followed by the name of the author who first described the species. Thus the mallard is known as *Anas platyrhyncos* Linnaeus.

One general principle of classification remains to be noted. We have so far tacitly assumed that the "characters" on which we divide groups are self-evident, but this is scarcely so. "Why," we may ask, "are whales classified as mammals rather than as fish?" Admittedly they have warm blood, mammary glands, and lungs—all of which are mammalian characters—but on the other hand they are aquatic, they have fins, they lack hair, and they have fishlike bodies. Why are their mammalian characters regarded as more important in classification than their fishlike characters? From evolutionary studies it has been shown that whales have developed from terrestrial mammalian ancestors, and their fishlike characters are merely adaptations superimposed upon their fundamentally mammalian bodies. The accepted basis for the selection of characters in classification is therefore the phylogeny (the evolutionary history of the race) to which the organisms belong. This is by no means the only possible basis, but it has been agreed by taxonomists to be the most meaningful and useful.

THE MAJOR GROUPS OF ANIMALS. Before we can discuss the history of life, it is essential to provide a brief outline of the more important divisions of the animal and plant kingdoms.

One of the most common expressions used to indicate the diversity of the scale of life is the phrase "amoeba to man." It is a familiar expression and readily brings to mind the complexity of some animals and the relative simplicity of others. In fact, however, the amoeba is a highly developed organism, which is almost as much more complex than the simplest organisms as man is more complex than amoeba. Below it on the scale of life come many other really simple forms. The flagellates, for example, are minute protozoans (unicellular organisms) some of which contain chlorophyll and yet are capable of the rapid movement one normally associates with animals. The Myzophyceae, the blue-green algae, commonly display no separation of the nucleus within the cytoplasm of their cells, and the bacteria, which resemble the blue-green algae in some respects, are even more primitive. Some diseases, however, including measles, influenza, and a number of plant diseases, are produced not by bacteria or other organisms, but by minute bodies of ultramicroscopic size known as viruses. These consist of nucleic acid and protein, and vary from about 10 millimicrons (ten millionths of a millimeter) to about 200 millimicrons in size. They pass the finest filters but are visible with the aid of an electron microscope, using magnifications of the order of 30,000. Their characters are such that they appear to stand very near the borderline of the world of living things. They bridge the size gap, for example, between molecules and the smallest undoubted living organisms (the smallest bacteria). They resemble living things in their reproduction within the body of an appropriate host, in their existence as definite strains, each with its own characteristics, and in their susceptibility to changes which are broadly similar to mutations. On the other hand, they can "exist" only in a living host, and their "reproduction" may be the result of the reproducing mechanism of the cell. They have not been shown to undergo any form of respiration, and, most remarkable of all, some of them may be crystallized and stored for a prolonged period, with no apparent loss of their characteristics; for if the crystals are afterwards introduced into the appropriate host, the virus will continue to develop.

These bodies, some composed of only a single molecule, are some of the most problematical structures known. Whether they are living cells, chemical entities, or complexes which assimilate the characteristics of living things from the host which they parasitize, but appear inanimate in the absence of a suitable host, we do not know. Indeed the viruses as a group exhibit sufficient variation in size to make it conceivable that they may include all three. They seem to lead us to the very threshold of life.

Viruses, like almost all the other really "simple" organisms, are not known as fossils, but they are a useful reminder that the history of life as we may read it from the fossil record is very incomplete. Below and within our level of observation and description there exist countless other organisms and processes which have played an obscure but vital part in the broader history of life.

The following list gives some indication of the character and content of each of the phyla.

KINGDOM ANIMALIA *(Animals)*

PROTOZOA — One-celled "animals," including the foraminifera and the radiolaria.

PORIFERA	Sponges, having porous body walls and commonly a siliceous or sometimes calcareous skeleton.
COELENTERATA	A single body cavity performs all the vital functions—the corals, jellyfish, sea anemones, etc.
BRYOZOA	Minute colonial animals with calcareous skeletons—the moss animals.
BRACHIOPODA	Bivalved marine shellfish—the lamp shells.
MOLLUSCA	Highly developed invertebrates, including clams, oysters, snails and slugs, nautilus, squids, etc.
ANNELIDA	Highly developed segmented worms—including a number of wormlike groups.
ARTHROPODA	Segmented animals with jointed appendages on each segment. Insects, crabs, lobsters, etc.
ECHINODERMATA	Spiny-skinned animals including starfish, sea urchins, sea lililes, etc.
CHORDATA	Animals with a notochord, including all vertebrates (fish, amphibia, reptiles, birds, mammals).

KINGDOM PLANTAE *(Plants)*

THALLOPHYTA	Simple plants, lacking roots, stems, and leaves, including bacteria, seaweeds, and fungi.
BRYOPHYTA	Plants with leafy stems: mosses and liverworts.
TRACHEOPHYTA	Vascular plants: ferns, trees, flowers, shrubs, grasses.

The Distribution of Life

One of the most striking facts that emerge from our discussion of living things is the way in which their structure adapts them to certain ways of life: the gills of a fish, the eggs of a reptile, the wings of a bird—all these allow their owners to live distinctive types of existence. But the corollary of this adaptation is that the more perfectly any organism is fitted for a particular mode of life, the less capable it is of surviving under different conditions. Pigs don't fly, and whales can't walk! Every living thing, therefore, has its own peculiar way of life, and this is controlled by its structure and physiology and by the nature of its environment. The relationship between an organism and its environment is a delicate and complex one, and its study forms the basis of the science of ecology (Greek *oikos,* dwelling). Wherever we travel on the earth today, we find living things—from the snows of the highest mountains to the waters of the deepest oceans, and under all kinds of conditions on, above, and within the earth. This great zone of life that envelops the earth is known as the biosphere and it includes three principal media, land, air and water, within each of which there exist countless smaller but distinctive environments. Thus on the land the animals of deserts are quite distinct from those of the frozen Arctic tundra or the tropical rain forest, and even within each of these environments there will be smaller areas and niches, each with its own individual physical conditions and its own peculiar fauna and flora.

But the teeming life of every nook and cranny of the earth is a thing of the present. Ancient organisms were much more restricted and the slow extension of the biosphere, as one environment after another was colonized, is one of the most dramatic aspects of the long history of life.

FACTORS CONTROLLING THE DISTRIBUTION OF LIFE. Although it is often easy to observe the general way of life and broad pattern of distribution of a given species, it is much more difficult to evaluate the various factors which influence its habits and control its range. Nor is this difficulty diminished by the fact that the factors which are of controlling importance in one environment (say a meadow) may be quite different from those which are of importance in another (such as the waters of a lagoon). Now what are these factors? We may think of them as being of three broad types, physical, chemical, and biological. The physical conditions will include such things as temperature, pressure, light and atmospheric conditions, depth of water, viscosity and diffusion of air, water currents, tides, topography, geographical position and composition of land surfaces, physical barriers to migration, conditions of sedimentation, and so on. Chemical factors include salinity, hydrogen ion concentration, gas and organic content of water, the oxygen, nitrogen, and carbon dioxide content of air, and the presence or concentration of a large number of compounds and trace elements. The external biological factors (as opposed to the "internal" factors represented by the organism's physiology and structure) are represented by the numbers and kinds of associated organisms, their interrelationships as food, prey, competitors, parasites, etc., mobility, population size and rate of change, birth and mortality rates of species, and various other factors.

Few of these factors are constant and most of them exhibit seasonal, diurnal, or random variation, which is often considerable. Clearly, therefore, the equilibrium that exists between organisms and their environment is both dynamic and intricate, and a slight variation in any one factor may have profound effects upon the general stability. There are various ways in which such environmental stimulus and change may affect individual organisms. They may, for example, be driven out of their original niche, they may be destroyed, or they may continue with or without some modification. These modifications themselves may also vary. The greater development of branches and leaves on the sunny side of a tree near a forest margin and the adjustment of the pupil of the eye to light of varying intensity are familiar examples of nongenetic adjustment known as acclimatization. Other changes favored by environmental pressure may be inherited in successive generations, however, and these are known as adaptations. The process of adaptation is a common one, and is of the highest importance in evolutionary development.

As C. H. Waddington stated in his article on evolution, "The argument from the fossil record" is "the strongest and most direct argument for evolution." A noted palaeontologist here introduces us to his field of specialization. His "long walk through Northwestern New Mexico" takes us through millions of years of geologic time and introduces us to the techniques by which fossils are made to reveal their histories. The present selection is taken from the author's well-known book Life of the Past. *In the Preface to another volume,* The Meaning of Evolution, *he compares the forest life of New Mexico with the human life of Hollywood, California. Both landscapes are "literally packed with life." They are "both parts of nature and both bear testimony to the wonder of life." Time has worked the miracle by which life was created from inorganic things, by which organisms of ever-increasing complexity arose, by which civilization itself developed from primitive beginnings. Yet all are interrelated, all are parts of a single whole.*

A WALK THROUGH TIME

GEORGE GAYLORD SIMPSON

LET US take a long walk in northwestern New Mexico. We may start from the rim of Chaco Canyon, where we can look down the vertical cliff and see a ruined pueblo, an ancient stone apartment house with 800 or more rooms. This one-house town was already old in the twelfth century but was flourishing then. Ahead of us to the northeast stretches undulating country, most of it pungent and gray-green with low sagebrush. Here and there along rocky crests are dark lines of bushy tree junipers and twisted pinyon-nut pines. On occasional rises along the sides of the broad, dry watercourses, the sandy arroyos of the Southwest, are areas of badlands. These are patches, from a few yards to many miles in extent, where the earth's crust is laid bare and has been carved by wind and rain into fantastic and austerely beautiful forms—wildernesses of twisted gorges, banded clay slopes, and weird sandstone-capped pillars.

We will seem to be solitary on our walk, striding across the vast, empty earth beneath the even vaster, incredibly blue, empty sky; but we will not be alone. Occasionally an antelope ground squirrel will dart across our trail or a coyote will slink away wraithlike at our approach. Always we will be watched by the alert dark eyes of Navajo Indians, crouching in the brush to guard their scattered flocks and aloofly curious about the actions of the white strangers.

A few miles from the canyon rim is a broad badlands area eroded in soft white and brown sandstone and gray clays with black lines that are the exposures of thin beds of coal. That coal, composed of ancient vegetation, reveals an exuberance of former plant life in strange contrast with the semidesert of today. A few minutes of prospecting along the clays reveal something stranger still: great bones much larger than those of any living American animal are being washed out by storms from their burial places in the hard clay. They are the remains of dinosaurs that swarmed here when the country was a lush lowland.

Another few miles and we come to a ledge of yellow sandstone in which are imbedded many large logs, fallen tree trunks feet in diameter and yards long. They look like fresh wood, inviting to the ax, but an ax swung against them would shatter its edge and strike sparks, for this wood is now silica, hard as rock crystal of which it is, indeed, a form. More extensive search shows that this sandstone, too, contains scattered bones of giant reptiles, dinosaurs known by such names as *Kritosaurus*, *Pentaceratops*, and *Alamosaurus*.

At the top of this sandstone there is a sharp line of division between it and a series of clays, banded in tones of gray with an occasional wine-red layer and lenses of white sand. Mark that line well, for it is the geological trace of one of the most dramatic events in the history of life. If we follow one of the wine-red bands a few feet above the division line and if we have the patience to keep our eyes glued to the ground for hour after hour and to crawl on hands and knees over the rough surface, eventually we will find ancient bones and teeth again. But what a difference! Here are no dinosaur bones. Most of these remains come from animals no larger than squirrels, and the largest of them might have to stretch to look a sheep in the face. If we study them closely we will find, too, that these bones and teeth are anatomically very different from those of dinosaurs. They are not remains of cold-blooded reptiles but of mammals, creatures warm-blooded like ourselves but smaller, vastly more ancient and more primitive than any human being.

We continue walking, still northeastward, and continue searching diligently for such remains of life as may be washing out from burial places in the successive clays and sandstones. By now, actually, our walk will have lasted for days, perhaps for weeks. We are covering only about thirty miles in an air line, but the ancient bones and teeth for which we are searching in successive layers are not abundant. At each level we must spend hours and often days to find any well enough preserved to tell us a clear story. With persistence we do find them, and we notice that their character is changing as we go along. No more dinosaurs are found, but the remains of mammals become more varied, some of the teeth become progressively more complicated in pattern, and larger bones do begin to appear.

We may stop finally among high mesas along the slopes of which are badlands magnificently banded in shades of red, lavender, yellow, and gray. Here we are almost sure to find, within a day or two, bones of an animal not unlike some we saw earlier in our walk but considerably larger and different in anatomical detail. *Coryphodon*, the giant of its time, could be compared for size with a particularly squat cow. Equally or more probable is the discovery of smaller remains quite unlike anything we have seen in the lower country behind us and of unusual interest to us. These belong to eohippus, the correct scientific name of which unfortunately happens to be *Hyracotherium*. A long sequence of discoveries elsewhere has shown that eohippus is, indeed, the "dawn horse," earliest known ancestor of Dobbin. With more time and luck we may even find a bit of the jaw of one of several sorts of animals smaller still and more interesting still: premonkeys, our own relatives some 60 million years removed.

This has been a walk through time. We have seen and touched a long segment of truly ancient history. Above the crust of the earth there have been tokens of shorter, human history: the pueblo ruin, the later Navajo invaders, and the latest invaders, represented by ourselves. This history is one of the outcomes of the longer history, but its few centuries pale to insignificance in comparison with the span we have followed in the exposed crust of the earth.

That span covered some 20 million years. It began toward the end of the Age of Reptiles while dinosaurs still ruled, perhaps 80 million years ago, and ended as the Age of Mammals was getting well underway, more or rather less than 60 million years ago. (I say "some," "perhaps," and "more or less" because these dates in years are not accurately determined.) The line of division between the yellowish, log-bearing sandstone, which geologists call the Ojo Alamo formation, and the overlying clays, the Naciмiento formation, was the line between the Age of Reptiles, the earth's Medieval Age or Mesozoic Era, and the Age of Mammals, the earth's Modern Age or Cenozoic Era. (See Figure V-2.)

At that line the last dinosaurs became extinct and the mammals, warm-blooded, furry, milk-giving, took over. At that time the mammals were still small, primitive, and not particularly varied, even though they already had a long history behind them. As we continued our walk we were witnesses to the expansion of the mammals, their progressive diversification, and the appearance of increasingly modern types.

The beginning of knowledge of the history of life comes from many, many walks like that by innumerable searchers in all parts of the world. The accumulated remains, with careful notes as to their places and sequence in the earth's crust, are sent to laboratories for preparation and study. They are identified and named. Their places in the past economy of nature are considered. Their relationships and lines of descent through the ages are determined. Little by little there emerges an ever clearer picture of what has happened in the history of life. More important still, we

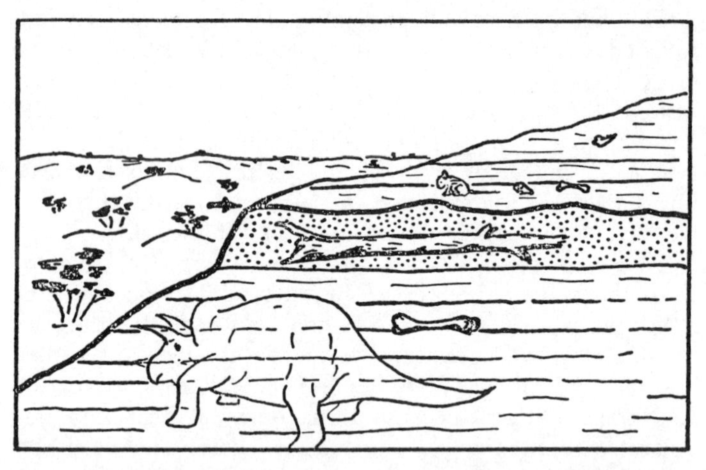

V-2. *Part of the rock succession in northwestern New Mexico. The lower rocks contain remains of dinosaurs, one of which is shown in ghostly restoration. The arrow points to the line between the Age of Reptiles and the Age of Mammals. Above this there are no more dinosaurs, but remains of small warm-blooded animals occur.*

begin also to see just how life has changed and to be able to judge why it has changed and why it is now what it is.

We are, ourselves, products of that history and we are its heirs. We cannot understand our own place in the universe or wisely guide our own affairs without knowledge of the processes that produced us and that still affect us and all the life around us. That fact alone makes knowledge of the principles of the history of life imperative for modern man, even if the history itself were not of intrinsic interest—and few who look into it fail to feel its fascination.

If you walk in New Mexico and pick up what is plainly a bone weathering out from ancient clays, or along the slopes of the Catskills and find what is plainly a seashell embedded in the rocks, you can say at once, "That is a fossil; it is a remnant of an animal that lived here long ago." The fact now seems quite obvious. It was far from obvious to many of our ancestors and was only generally accepted as a fact after centuries of learned argument. The long history is a good example of how painful and groping is man's rise to knowledge and how faith, dogma, and authority can make us blind to the plain evidence of our senses.

As early as the sixth century before Christ the Greeks knew in a general way what fossils were and what they mean, but as late as the eighteenth century of our era, some 2200 years later, men of science were still gravely arguing the point. A long list of Greek philosophers and historians, among them Xenophanes, Xanthos, and Herodotus, noticed that seashells may be found buried far inland and concluded that the sea had once stood where the shells were found. Bones of mammoths were also known to the ancients. They recognized these as bones but usually ascribed them to gigantic men—an interpretation still generally accepted in the eighteenth century and occasionally thereafter.

During the Dark and Middle Ages the ancient lore about fossils, rudimentary as it was, was largely forgotten. Among those who paid any attention to fossils at all the most popular theories were that they had been engendered in the rocks by a sort of "formative force," or that they had grown in the ground from germs fallen from the stars, or in some cases had fallen fully formed from the heavens. No clear distinction was made between what we now call fossils, that is, the actual remains or traces of ancient living things, and such things as crystals, old stone axes, or rocks that happen to have the shape of an ear or some other organ or animal. Even in the sixteenth century, when Agricola coined the term "fossil" (from Latin *fossilis*, "dug up"), it meant any curious object found buried. Figure V-3

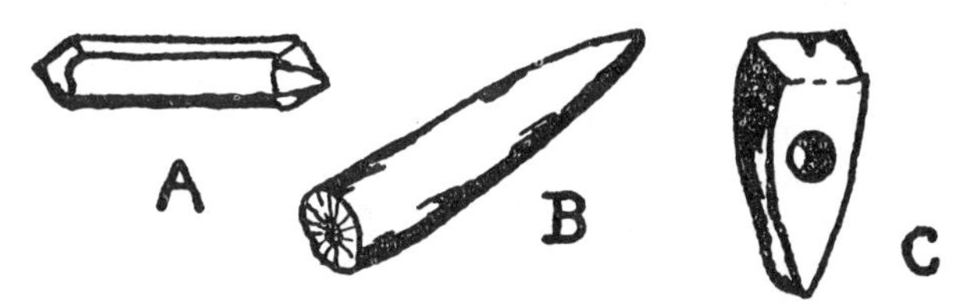

V-3. *Some of Gesner's fossils (1565). A, a quartz crystal. B, a belemnite, part of an extinct relative of the cuttlefishes. C, a prehistoric stone ax. Only B would be called a fossil today.*

shows some of the things that Gesner called fossils in 1565. Only gradually thereafter did the word come to be confined to traces of life, the only usage now current.

While most scholars were arguing fluently and almost entirely erroneously as to the nature of what we now call fossils, an opposite error was becoming widely accepted. In the thirteenth century Albertus Magnus was probably echoing general opinion when he said that "Whole animals can be petrified. The constituents of the body of the animals are modified. Earth mixes with water and mineral matter changes the whole thing to stone while preserving the form of the animal." We will see below that something rather like this can happen but that it is extremely rare in animal fossils. The objects to which Albertus was referring were almost certainly not petrified animals but merely stones with an accidental resemblance, like the "ear stones" mentioned above. This idea dies hard, as all mistakes seem to. Museums are still being bombarded with "petrifactions," the only connection of which with a fossil animal is in the eye of the beholder. Only the other day I was offered for sale at a large price "the petrified leg of a woman." I was called a liar and a cheat when I explained that it was only a piece of volcanic rock with an accidental (and very slight) resemblance to the vision in the mind of its owner.

A few of the learned did retain the ancient idea that fossil shells were shells in fact and that where they occur the sea has been. To be learned then was to be a cleric, and it was inevitable that sooner or later someone would have the wonderful idea that fossils are witnesses to the Biblical deluge. Ristoro d'Arezzo, an Italian monk of the thirteenth century, is sometimes credited with this inspiration. Whether he was first or not, he did express the idea in a work on the "Composition of the World" in A.D. 1282. This error, too, has never died out altogether. It is still taught as, literally, gospel truth in some church-controlled schools in the United States. In the intellectual atmosphere of the Middle Ages, which these anachronistic schools still breathe, it is not surprising that this soon became the predominant theory as to fossils.

Leonardo da Vinci in the late fifteenth century was a precursor of modern geology as of so much else that is modern. He rejected alike the views that fossils are mere "plays of nature" and that they are testimony to the deluge. He said, "The mountains where there are shells were formerly shores beaten by waves, and since then they have been elevated to the heights we see today," and he produced logical, keenly observed evidence for this correct opinion.

Thereafter interest in fossils was more intense and continuous than before, but for two centuries this meant that agreement as to their significance was even less general and disputes were even more bitter. If space permitted these centuries would yield a noble roster of honest observation and logical deduction, an infamous roster of educated bigotry, and a ridiculous roster of superstition and surmise. High on the first list belongs, for instance, Bernard Palissy, a French pottery maker and self-educated naturalist. He collected fossils and concluded that they were exactly what we now know them to be. He even correctly identified the species of some of them in a thoroughly scientific way, probably the first time that this was ever done. Finally he went to Paris and gave a series of lectures attacking the dignitaries of the Sorbonne, whose opposition won them places well up among the educated bigots. Palissy was right, but the Sorbonne won: Palissy died in the Bastille in 1590.

Nearly a century after Palissy's ill-omened debate a Dane, Nicolaus Steno, living in Florence, published a book dated 1669 in which he not only recognized the true nature of fossils but also pointed out the successional nature of rock strata. The works of Palissy and of Steno, between them, contain all the essentials on which a true science of palaeontology, of fossils and the history of life, could have been and was, indeed, later to be based. Yet it was again well over a century after Steno until it was definitely established and generally agreed that there is a succession of different fossils in the rocks and that this succession is the record of the history of life. Of course others contributed greatly, before and after, but this final triumph was largely due to William Smith (1769–1839) in England, G. B.

Brocchi (1772–1826) in Italy, and Alexandre Brongniart (1770–1847) in France.

Even so brief an account must not be ended with the impression that history moves steadily forward, that knowledge and understanding progressed from da Vinci to Palissy to Steno to Smith and onward to us. A few brief indications to the contrary: In 1699 Edward Lwhyd published what was for the time a magnificent atlas of fossils and explained to the reader, as medieval monks had said long before, that these were engendered in the rocks from seeds carried by wind and water. In 1695 John Woodward used his fine collection of fossils as a basis for "an Account of the Universal Deluge." J. J. Scheuchzer, Swiss, was so delighted with Woodward's book that he followed it by several of his own, including one that proved from fossii evidence that the deluge occurred in the month of May. (This conflicted with the previous demonstration by an English astronomer, William Whiston, that the flood occurred on a Wednesday, November 28.) In 1746 Voltaire, a less successful universal genius than da Vinci, wrote, "Is it really sure that the soil of the earth cannot give birth to fossils? A tree has not produced the agate which perfectly portrays a tree. Similarly, the sea may not have produced these fossil shells which look like the homes of little marine animals." And in 1952 there is a purported textbook of geology on the market that would have seemed laughably antiquated to Voltaire.

In spite of the fact that old errors never die, it has been general knowledge among competent students for about two centuries now that fossils are the remains of ancient organisms, that they can be identified and classified by anatomical comparison, that they occur in sequence in certain rocks of the earth's crust, and that they form a historical record. There has been much more recent dispute as to the meaning of that record, but that is a different point. Now let us consider what sort of things fossils are.

How wonderful it is that organisms so ancient can be preserved for us to see, handle, and study! Plants and animals die around us today, and we usually find no recognizable trace of them a few years, a few months, sometimes even a few hours later. Yet some fossil plants are well preserved after more than a thousand million years, and the evanescent delicacy of a jellyfish has lasted as much as 500 million years.

Preservation of an animal entire, just as it died, is the rarest of accidents and with a few exceptions has persisted only for organisms so recent as hardly to merit designation as fossils. They are called fossils principally because they are of species that became extinct in prehistoric times. Any remains really ancient, from, say, 100 thousand years upward, are agreed to be fossils no matter how they may be preserved. Younger fossils, on into the dawn of history, intergrade with recent remains and prevent hard and fast definition. There is no exact point when an animal becomes a fossil. Certainly this is not when it petrifies—most fossils never do really petrify.

Extinct animals preserved whole, or nearly so, are then a sort of subfossil. They include the famous mammoths found in Siberia, frozen whole and preserved in cold storage for some thousands or perhaps at most tens of thousands of years. In Alaska, too, considerable parts of mammoths and of some other animals have been preserved in frozen muck. Mammoth hair is sometimes common enough there to be a nuisance in the gold diggings. In Starunia, Galicia, a young mammoth and a young woolly rhinoceros were found in a deposit of waxy hydrocarbon which had preserved them nearly whole. In particularly dry caves in southwestern United States and in Patagonia were found ground sloths perhaps a few thousand years old, not whole, to be sure, but with large parts of desiccated hide and hair, tendons, and piles of excrement preserved in addition to the bones.

Interesting as they are, such finds are so rare that they are more curiosities than usual objects of palaeontological study. More common and of more scientific importance are the occurrences of parts of plants and of partial or whole animals, mostly insects, in amber. Amber is the hardened, chemically altered resin of ancient trees. When the resin was soft many insects and a few other organisms were trapped and sealed in it, embedded in a tomb of transparent, antiseptic, natural plastic. The famous Baltic amber used in jewelry from ancient times is particularly rich in beautifully preserved whole insects. Because the amber where found is not in its original deposit but has been washed in from somewhere else, its exact age is unknown. It may be on the order of 50 million years old, and amber with insects of possibly still greater age is known from other regions.

There are a few other antiseptic burial places where soft tissues may be preserved with little change. Noteworthy among these are bogs. Soft coals formed in ancient bogs may be tens of millions of years old, as in Victoria, Australia, and contain wood that is somewhat darkened but otherwise so unaltered that it can easily be cut with a saw and planed. In Germany a similar deposit has produced remains of animals badly flattened and distorted but with some tissues so fresh that details of soft cells can be seen under the microscope. The bottom waters of some lakes and arms of the sea can also become antiseptic with accumulation of chemicals so that decay does not occur there. To this circumstance we owe such extraordinary finds as fishes, also flattened and blackened, but so preserved that after nearly 300 million years the muscle fibers and their cross-striations are still visible with a microscope.

Practically speaking, truly petrified animals with the whole body, soft parts and all, turned to stone do not occur, in spite of Albertus Magnus and some enthusiastic but ill-informed moderns. Perhaps the nearest thing to such preservation is that of whole bodies of frogs and a few other animals replaced by phosphate minerals near Quercy in France. The occurrence is altogether exceptional, indeed unique.

The preservation of wholly soft animals without skeletons and of soft parts of animals with skeletons is not very rare, but it commonly occurs in two ways only: as thin films of carbon and as plastic impressions on the rocks. All soft tissues contain carbon in complex compounds. In the processes of partial decay and burial, with increasing pressure as fossil-bearing rocks accumulate, these compounds may break down in such a way that other volatile or soluble materials escape but a thin film of black carbon is left. That is the way the 500-million-year jellyfish was preserved, and with it a fascinating array of other soft-bodied creatures. Most of these were collected in British Columbia, Canada, and described by the late American palaeontologist C. D. Walcott. It is also the explanation of the preserved body outlines of the famous ichthyosaurs from Holzmaden, Germany, but even there this is a rarity. Most of the ichthyosaurs are preserved as skeletons only. Preservation of fossil leaves as carbon films is very common, indeed usual, but there is usually a plastic impression also.

The rocks in which fossils occur were almost always soft sediments, muds, silts, and sands, when the remains were buried. Even the most delicate tissues may make an imprint on such sediments, and if another layer is deposited without destroying the imprint it may be preserved as the rock hardens and may last indefinitely. Fossil jellyfish have been recorded in this way, too. Or the body may decay after burial and leave a cavity, which may remain as a hollow in the rock or be filled by sediment or mineral later on. One of the most remarkable instances was recently found in Oregon. There an ancient rhinoceros, apparently already dead but with the skin still intact, was overwhelmed by a lava flow that quickly hardened around its body. What remains now is a cavity reproducing the shape of the bloated animal, and inside the cavity were a few fragments of burned bones.

The famous dinosaur "mummies" are not in the usual sense mummies. After the animals died the skin dried over the skeleton and the whole thing was then buried in sand. The actual skin and all the soft internal organs have long since disappeared, but clear impressions of skin are preserved on the sand now turned to hard sandstone.

All these ways of preservation are interesting and enlightening when they occur, but at least 99% of all animal fossils, to be very conservative, and a great many plant fossils have the skeletons or hard parts preserved and no others. For the vertebrates what is usually preserved is the hard bone and the still harder teeth, nothing else. Sometimes cartilage is so impregnated with lime in living animals that it is as hard as bone and may be as well preserved. We used to speak of fossil bone as petrified and

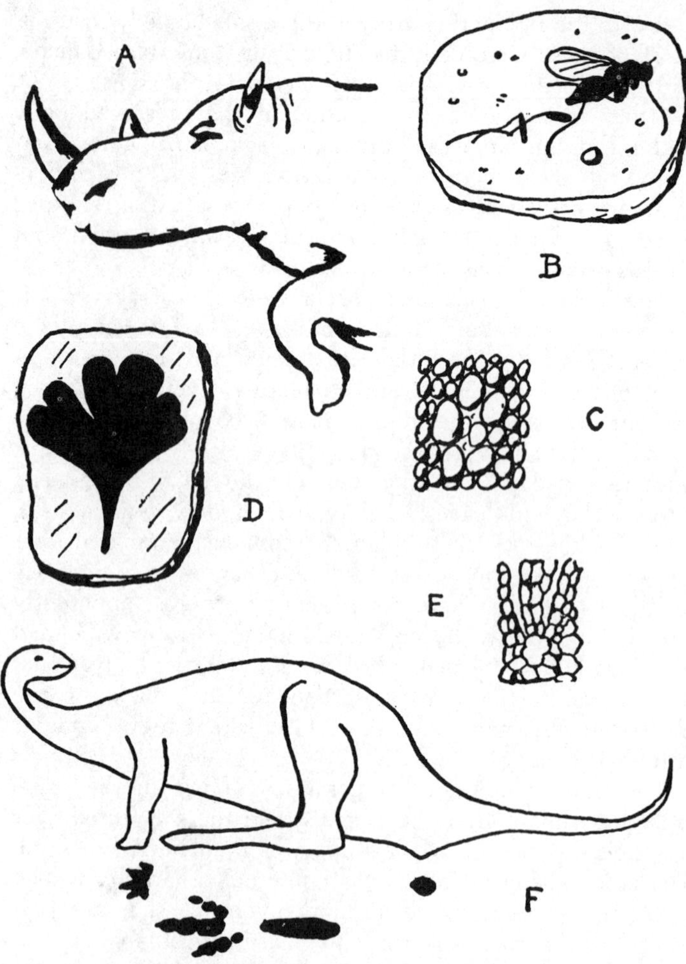

V-4. Some ways in which fossils are preserved. A, front end of an extinct rhinoceros, preserved entire in mineral wax (after a photograph by Niezabitovski). B, a fly in amber (after a photograph by Bachofen-Echt). C, impression of dinosaur (hadrosaur) skin in sandstone. D, leaf (ginkgo) preserved as a film of carbon. E, petrified wood, greatly enlarged thin section showing microscopic cell structure. F, footprints and rump-print of a sitting dinosaur (Triassic of the Connecticut valley) with sketch restoration of the primitive dinosaur in the position in which the prints were made (data from Lull).

still do sometimes in a loose sense. We now know, however, that it is seldom petrified as our ancestors understood that word and most nonpalaeontologists still do. Even in the oldest fossils the original bone substance has seldom "turned to rock," or even been replaced by some quite different mineral. Usually the original hard material of bone and teeth which formed when the animal was alive is still there. Perhaps it is somewhat rearranged in structure but with little or no change in composition, a complex compound of lime and phosphate. The soft materials that occupy larger or smaller cavities and canals in the hard bone or tooth tissue soon decay and leave empty spaces. Sometimes that is all that occurs. More often these spaces are later filled by some mineral—silica (silicon dioxide, the substance of rock crystal or of many sands) or calcium carbonate (the substance of limestone) are the most common fillers but many others may occur. It is this filling that makes fossil bone heavier than recent bone and also more brittle. Usually the original bone or tooth substance becomes discolored. Fossil bones and teeth are rarely white, often black, and may be almost any color of the rainbow.

Most shells are made of some form of calcium carbonate, and this, too, may be preserved in fossils without change or with only microscopic recrystallization. More commonly than in the case of bones, however, this rather soluble material may be leached out, leaving a cavity in the rock, or may be replaced by some other mineral. Replacement by silica, a much harder and less soluble material than calcium carbonate, is frequent and has special value for preservation and recovery of the fossils. Some shells and the outer coatings of all crustaceans, insects, spiders, and their relatives are composed of chitin or organic compounds of similar properties, flexible but tough, something like your fingernails although of different chemical composition. Such materials are very resistant to decay and may be preserved indefinitely in fossils without marked change.

Internal cavities, such as occur in shells or in skulls, are often filled with matrix (the sediment in which the fossil is buried) or with some mineral and may be preserved

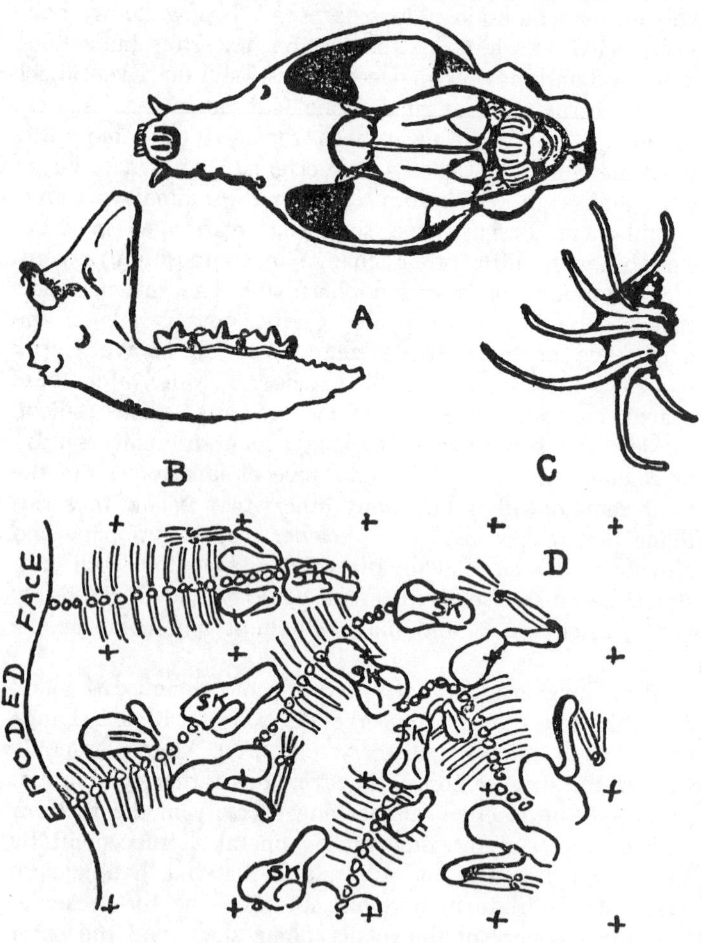

V-5. *Some kinds of fossils. A, skull of an early carnivore (Cynohyaenodon, early Oligocene) seen from above; the top of the brain case is broken away and a natural cast of its interior is exposed. B, broken lower jaw of a fossil shrew (Domnina, early Oligocene), more typical than A or D of usual materials for the study of fossil mammals; the original is less than half an inch long. C, a fossil sea snail (Pterocera, Jurassic), the shell preserved whole, as is frequent for fossil shells. D, rough quarry diagram of part of a deposit of skeletons of an early mammal (Coryphodon, early Eocene), marked in two-foot quadrants; six skulls (SK) and parts of associated skeletons are seen.*

after the surrounding shell or bone is dissolved or weathered away. These internal casts, often called endocasts for short, are also valuable fossils which reproduce the gross forms of the soft parts that filled the cavities when the animals were alive.

The outermost tissues of plants often contain a hard waxy substance, cutin, which is resistant to decay and may preserve microscopic detail in fossils long after other tissues are carbonized or entirely lost. The woody (cellulose) tissues of plants are frequently very well preserved, even to minute features of the cells, by what is still usually called petrification. It has, however, been found, as it was for bones, that the process does not correspond with old and popular ideas of petrification. What happens first, as in the case of bones, is simply that the hollow spaces and those left by decay of soft organic materials are filled by a mineral deposit. Then the cellulose walls may carbonize or decay and their places may be filled by a slightly different mineral, but some of the organic material of the walls commonly remains even in ancient fossil plants. As with replacement and secondary filling in animal fossils, the minerals concerned are usually calcium carbonate or, especially, silica. Many other minerals may be involved, but most petrified wood is silicified.

Besides their actual remains, ancient animals have left other traces that are preserved in the rocks and are also considered fossils. The dinosaur eggs found in Mongolia are famous, and other fossil reptile and bird eggs have been discovered. Although fascinating, they are of minor scientific importance because their occurrence is so sporadic and they tell so little about the animals that laid them. The same is true of gizzard stones or gastroliths, gravel or pebbles swallowed by reptiles and birds to help grind their food. Moreover, most of the polished "dinosaur gastroliths" cherished by amateur collectors were never really inside a dinosaur or any other animal. Oddly enough, the excrement of some animals, especially carnivorous animals, fossilizes readily into a hard phosphatic mineral mass, and fossil pieces of excrement, called coprolites, are abundant. The burrows of animals of many sorts are also fairly common fossils, formed when the burrow is filled by some sediment that contrasts with their walls.

Tracks and trails of all kinds of animals, from worms to men, may also be preserved in the rocks. They are widespread and important fossils known almost throughout the long fossil record. The tracks called *Chirotherium* have a long and curious history, and dinosaur tracks from many parts of the world are famous. Especially diverting are footprints found seventy years ago in hard rock in the courtyard of a prison at Carson City, Nevada, and long popularly interpreted as human. Mark Twain wrote a humorous article making them the aftermath of a drinking bout. They finally turned out to be tracks of an extinct ground sloth.

There are still other sorts of fossils, but enough has been said to suggest the usual nature and great variety of these extraordinary messages from the past.

To those who follow it, the pursuit of fossils is more exciting and rewarding than any pursuit of living fish, flesh, or fowl. It has all the elements of skill, endurance, suspense, and surprise; and the resulting trophy may be a creature never before seen by man. To be sure, fossils do not fight back—they do often seem to elude the pursuer—but I have never seen a fox riding to hounds or a lion carrying a rifle.

The question most often asked of a fossil collector, especially of the variety "bonedigger," is "How do you know where to dig?" There are so many different kinds of fossils and ways to hunt for them that a complete answer would require several books this size. The truest short answer is "You don't." One usual way to hunt for fossils follows the procedure you would have to follow if you were told to go find an object an inch long or less, exact shape and nature unknown, supposed possibly to have been dropped somewhere in an area of wilderness ten miles square. There is no esoteric sense and no instrument to tell that a fossil is buried at a given place. The collector just has to go look until he finds fragments of fossils exposed at the surface by erosion. If the fossil has entirely weathered out he picks it up. If part of it or if other fossils are still embedded there then he knows where to dig, and does so.

Although the procedure is often no more than patient search, special skill is involved. The hunter does not know beforehand where to dig, but he needs to have a good idea of likely areas in which to look for places to dig. Without knowing just what he is going to find, he has to recognize it when he does find it. Local inhabitants who are not amateurs of palaeontology often walk over fossils every day of their lives without seeing them. And a fossil in the field does not look like one in a museum case. All that signals to the eye "fossil" and not "rock" may be an indefinable difference in color or the slightest oddity of shape. In a different kind of fossil hunting, fossils may even have to be collected without being seen at all. The collector has to judge what bed of rock might contain microscopic fossils and then take a sample of it back to the laboratory, perhaps a thousand miles or more away, before finding out whether he is right.

In any case a good fossil hunter must be a skilled geologist. For modern scientific purposes a fossil has no value unless its exact place in the sequence of rocks is determined. Usually, too, information as to the nature of the rock, its manner of deposition, and other geological details are required.

There is also skill in extracting fossils from their burial places without damage. There are many techniques for different circumstances, among which the most elaborate is perhaps that used in modern collecting for taking out large, brittle, and cracked skeletons. This involves careful exposure of the deposit from above, separation of the bone-bearing bed into blocks of manageable size, hardening the bones and matrix with shellac or plastics, and encasing each block in splinted plaster and bandages. The procedure is more complex than suggested by this brief summary. Experience and usually a rather elaborate set of special tools and materials are necessary for success.

Some fossils are weathered out or can be broken out in the field in such shape as to be studied without further preparation except, perhaps, soaking with a preservative.

Most of them, however, require further cleaning and piecing together before all their preserved features can be well seen. In the laboratory, too, there are innumerable different techniques adapted to different kinds of fossils and the different sorts of rock in which they occur. Only a few examples can be mentioned in this brief account. Sometimes removal from the rock is best accomplished by long labor with hammer and chisel. Dental drills and other small rotary grinders or pneumatic chisels are often useful, or small fossils may be scratched out with awls and needles under a microscope.

Of great importance in recent studies of fossil shells, especially, are methods of chemical preparation. It is often possible to find parts of a limestone bed in which fossil shells are abundant and have been replaced by silica. If blocks of such rock are soaked in hydrochloric acid with appropriate precautions, the rock itself is dissolved and the fossils remain. Extraordinarily delicate features, such as hairlike spines, may be preserved by this method although lost or not even noticed if preparation is mechanical rather than chemical. Even more important for modern study of communities and populations of fossils is the fact that the method facilitates mass collecting. A large mass of rock treated chemically may produce thousands or even tens of thousands of separate fossils. Other chemical methods are also used. For instance chitinous fossils can be freed from a silica matrix with hydrofluoric acid.

In addition to cleaning fossils from the rock in which they are embedded, some groups of fossils and some sorts of studies require a large variety of additional methods. Fossil wood and some small animal fossils can be precisely identified only by means of the microscopic internal structure. This can be seen in slices so thin as to be transparent, slices that cannot be cut to this thinness in such hard material and must be ground. Sometimes an equivalent result can be obtained by polishing a surface on machinery similar to that used in gem cutting. The polished surface is then etched with acid and painted with certain plastics. When dry the plastics can be peeled off, and the peel retains an imprint of the fine structure of the fossil. Peels can also be used to follow structure all the way through a fossil such as a small skull. After each peel is pulled the surface is ground down again and the process repeated at very short intervals. From these serial peels, as they are called, enlarged wax or plaster models of the original fossil and its internal features can be made. There are also special ways of making casts of internal cavities. They can be filled with some acid-resistant material and their walls then removed by acid. Or, especially for study of brains and associated nerve patterns, a skull may be cleaned out and the inside painted with latex, which hardens to form but remains flexible enough to be pulled out in one piece.

Staining of fossil tissues, radiography and ultraviolet photography, rotary milling of matrix, flotation on heavy liquids, and many other methods are also used for study and recovery of fossils of various sorts. Over the years it was learned what fossils are and how to find and collect them. Now we are learning how to extract from them all the information they can give us, which is much more extensive than our predecessors dreamed.

In a small book entitled The Living Past, *John C. Merriam, the American palaeontologist who was President of the Carnegie Institution of Washington from 1920 to 1938, described the pleasures and adventures of fossil hunting. He recounted the discovery of the La Brea tar pit in California, in which such extinct animals as saber-toothed tigers had been trapped, the death of an Indian maiden in a hidden cave, and the giant redwoods which are "a living link in history." One of his most interesting tales is reprinted below. It relates the astonishing discovery of a fragile leaf which had fallen from a tree and been preserved over a period of hundreds of centuries.*

THE STORY OF A LEAF

JOHN C. MERRIAM

IT HAPPENS often that the greater things of nature, as in human life, are only imperfectly appreciated until in the light of a single circumstance suddenly they arise to overshadowing importance. So we read that when Elijah stood upon the mount before the Lord, it was not in the strong wind, nor yet in the earthquake or the fire, but through the "still small voice" that the Almighty was recognized.

On a recent journey to Columbia River Gorge I had the pleasure of revisiting a region familiar from many years of interest in the meaning of its history. As on other occasions, the bordering cliffs, with summits disappearing in mist and the river moving majestically below, made me see the canyon in all its strength and beauty. In the geologists' language, I read again the record of ancient floods of lava piled up to build a mountain in which later the river cut this gorge. But it remained for a fragment of a leaf, fluttering from its burial place in the foundations of the valley wall, to give through compelling vividness of its own reality the clearest expression of this story of creation. The finding of the leaf continues always fresh in memory as one of those occasions when the past seems to open, and for the moment we look through to see the Builder at work.

It was on a stormy day in July that I passed through Columbia Gorge with a party of geologists engaged in study of the history of that fascinating region. Riding down over swinging curves of the highway we examined the rock formations which make the canyon rim. The summit was formed by edges of old lava flows. Below the lavas were thick layers of water-washed gravel and boulders. Deeper in the gorge the gravels lay upon still greater lava flows, piled hundreds upon hundreds of feet to make the wall.

Across the river these lava beds were seen extending widely under mountains and plains beyond.

As the panorama lengthened our wonder deepened. Rising two thousand feet around and above us were stony cliffs of what was once melted rock. Below, like a great ribbon-saw, the river was cutting its way with teeth of sand and gravel dragged upon its floor. We recognized the canyon as merely one stage in a process of shaping and molding of the vast pile into which it had been cut.

Approaching the historic station of Bonneville, our party found the lower beds of lava along the highway resting upon a foundation of rocks that had formed the landscape over which these first flows were poured. This underlying formation consisted of layers of hardened sand, gravel, and mud. A member of the party who had made special study of the region stated that at this point the rocks beneath the lava contained buried in them remains of plants that flourished here before the time of the first lava flows.

It was where the highway skirts a steep bluff, before bending round the curve at Tanner Creek, that search was made for the locality reported by the specialist in history of plants. The cliff rose almost perpendicularly from the pavement. Its irregular face was marked by widely spaced bands of stratification representing difterence in materials. In the first large exposure bordering the road, a log of petrified wood projected from a rock which had once been the mud and sand that covered it. A few yards away the twisted stem of a dead tree of modern time hung over the cliff. The fringe of forest, draping the bluff above, was formed of fir, maple, willow, and dogwood. Near the fossil log a living alder clung to the face of the rock, and a sword-fern had fastened its roots in a softer stratum close by.

Rapid search revealed the locality reported to furnish remains of plants. A few strokes with a pick in the soft layer from which the sword-fern grew brought out fragments of rock covered with clean-cut impressions of leaves. There were maple, sweet gum, hickory, and oak, with numerous other kinds. The palaeobotanist, who had spent many days collecting in this stratum, began to name the plants discovered. I interrupted to ask if he had obtained the Gingko, a beautiful tree originally known living only in groves about the temples of China and Japan, but found also as fossils at localities spread widely over the northern half of the world. "Yes, I have found it, but only rarely," was the reply. "Oh," said a student standing near, "I have just seen an impression of a leaf." He ran to the cliff and brought a small slab. On one edge was a narrow imprint showing clear tracery of a leaf like a maiden-hair fern, but duplicated only by the Gingko.

When the print appeared my mind went back to a tree of this strange type planted in the court of the Cosmos Club in Washington. I remembered sitting in the yard at lunch on a warm summer day and pulling down a leaf as reminder of the quest I wished to see for wild Gingkos in primitive forests of Asia. I thought of the history of the tree as known to the palaeobotanist—for millions of years widely spread over the northern hemisphere, and now only a slender residue remaining protected in the sacred forests.

Still looking at the slab, but with my thoughts in Washington, in China, and in the world of past ages, my youngest son brought me back to Columbia Gorge by remarking that he saw the leaf itself still in the rock. "No," I replied. "It is only the impression that remains. In these formations the leaf decays and disappears. Not even a film of coal is left from the carbon which it originally contained." "But the leaf is there," he said. Then, to prove the truth of my statement, taking the specimen, I pulled apart the layers on which the impression was made, and as they separated, clinging to the print, we saw a brown fragment of a leaf with one edge lifted and moving in the wind. The structure was that of a Gingko, time-bronzed and shriveled—but a leaf, and not merely the trace of its form upon the rock.

My eyes turned to the bluff from which the fragment came. Beyond the layers of sand and mud in which the leaf had been entombed rose cliffs of lava, stepping up and up to the rim of the gorge. I looked across the chasm the river had cut, and then saw again the remnant of the Gingko fluttering in the breeze. Since last the wind had stirred this fragile thing, a mountain had piled itself upon it. Floods of melted rock had sealed it in. And then the river, sawing for ages through the mass, had made easy its liberation by the hammer of a student seeking to know the truth about the past.

In careful wrappings the Gingko fragment from Columbia Gorge crossed the continent to tell still further details of its history through microscopic examination of its structure. At the laboratory in Washington it was lifted from the rock to which it adhered, and after careful preparation came to the table of the microscope. By its side lay a similar portion of a Gingko leaf from the tree in the courtyard of the Cosmos Club.

The expert first examined the modern leaf, noting the patterns of its cells. Then the ancient fragment took its place in the field of magnification. The lighter tint of the living leaf was replaced by brown of the specimen from beneath the lavas, but pattern of leaf and form of cell were as much alike as two closely related ferns or oaks of today.

As he moved it about under the microscope, the specialist ran his eye rapidly over the surface of the ancient leaf-blade. He had hoped to find all details of structure present, but the guard cells bordering the breathing pores were not to be seen. He ventured the suggestion that "perhaps after all this never was a living, growing thing, but only an accidental resemblance to a leaf, or an incomplete experiment in creation." "Possibly," I replied, "you are looking at the wrong side of the leaf." The fragment was reversed. A few quick shifts followed, as the object was moved across the field of vision. Suddenly there was close examination of something that caught his eye. He changed the focus, and then motioned me to his seat at the microscope.

Looking down on the illuminated leaf, I saw dotting its surface characteristic pairs of cells which guard the pores through which plants breathe. It was clear that in this specimen, as in the green Gingko fragment from the living tree, these little mouths had opened and closed through the rounds of daily life for a season. And then on a late summer day, as a golden leaf with its work completed, it had whirled down to the stream that buried it.

For a moment the fragment faded from clear focus, as I remembered the leaf where it lay beneath the wall of the canyon in layers of rock that had been mud washed upon it ages before. And then the eye of my mind did what as yet no man-made instrument has accomplished. Through the assured reality of this leaf, as it was then and had been so long ago, a view was opened across the eons connecting present and past. In that sweep of vision it was not possible to avoid the panorama of history just as it had occurred. Glowing lavas crept out, flow on flow, through ages to build the mountain; and then a river began its work of fashioning the canyon.—Suddenly my friend inquired: Had I seen the guard-cells? "Yes, and also traveled out across the continent, and back through time so far I might not guess the distance."

The cliffs in which the Gingko lay entombed in Columbia Gorge are included in Mount Hood National Forest. Some years ago a portion of the area was set aside by the Forest Service as Columbia Gorge Park. Within this region, by help of cards of description, the traveler is assisted in his effort to know the lessons of the place more intimately. When next you visit the gorge, approaching Tanner Creek on the highway from the west, you will find a simple marker on the cliff in which the Gingko leaf was found, and in which others still rest. At the foot of the wall you will see a Gingko tree planted from stock that grew about the temples of Japan. Its branches brush the rocks in which the ancient leaves are buried.

If you frequent the canyon, you will find that no matter how often you follow the highway from Portland to the Columbia, there is always a thrill when the ridge above Crown Point is crossed and suddenly the gorge appears. It may be that to some this panorama expresses itself solely in terms of superficial form and beauty. Looking often over the valley, it has seemed to me that no one could view this picture without finding revealed something of that movement of creation of which it is the result. Commanding as are the features of external beauty, evidences of the process of making portrayed are of such significance that details of the spectacle must be recognized as only incidents in a moving panorama. To know the whole story takes nothing from appreciation of form and feature as they appear at the moment. This knowledge only extends our vision. We come to realization that the elements of grandeur comprised in bulk and static strength have their present values because they are residual demonstrations of power exerted in a great work of construction.

Today we begin to see through all nature the movement of great forces involved in the process of creation. As this panorama opens we glory in our growing power of comprehension. Then on a rare occasion, in an unexpected region, there appears a new expression of reality far transcending that built upon previous experience and imagination.

This flash of understanding, with widening horizon, can come to each individual as veritable revelation. It may bring realization of unmeasured power bound in the atom, another universe beyond the stars, a new vista in the past, or an explanation of some baffling phase in human life.

We may stand often, as it were upon a mountain, with the elements spreading their power before us, and yet miss the gentler message that makes it possible to see behind the face of nature and read in some small part its deeper meaning. It was under such circumstances that a fragment of a leaf opened the doors upon a living past. There fire, earthquake, and the flowing power that cleft the mountain, each was seen to have its part in building order and beauty into the world in which our lot is cast.

The study of living things in relation to their environment —their habits, life cycles, populations, interrelationships—is the modern ecologist's domain. In its very nature, ecology is not now and may never become an exact mathematical science; yet its practical importance cannot be overestimated. The importation of rabbits into Australia, their multiplication and the resultant effects on vegetation are a proper study of the ecologist. The eradication of anopheles mosquitoes and thus of the malarial bacteria which they transmit is a medical-ecological problem. Silent Spring, *the best-selling book by Rachel Carson, a discussion of the effects of insecticides, is an ecological document. Indeed, the myriad ways in which man has upset the balance of nature and drastically altered the environment is perhaps the most frightening aspect of ecological research. Only in the twentieth century has the importance of such study been fully appreciated, and the formulation of exact methods been attempted. One of the leading twentieth-century ecologists was Charles Elton, the British biologist whose book* Animal Ecology, *published in 1926, formulated some of the basic concepts of modern ecology. From it has been selected the following chapter, which describes some of the self-regulatory measures that nature has devised for animal communities.*

THE ANIMAL COMMUNITY

CHARLES ELTON

"The large fish eat the small fish; the small fish eat the water insects; the water insects eat plants and mud."
"Large fowl cannot eat small grain."
"One hill cannot shelter two tigers."—CHINESE PROVERBS.

EVERY ANIMAL is (1, 2) closely linked with a number of other animals living around it, and these relations in an animal community are largely food relations. (3) Man himself is in the center of such an animal community, as is shown by his relations to plague-carrying rats and (4) to malaria or the diseases of his domestic animals, *e.g.*, liver-rot in sheep. (5) The dependence of man upon other animals is best shown when he invades and upsets the animal communities of a new country, *e.g.*, the white man in Hawaii. (6) These interrelations between animals appear fearfully complex at first sight, but are less difficult to study if the following four principles are realized. (7) The first is that of *Food chains* and the *Food cycle.* Food is one of the most important factors in the life of animals, and in most communities (8) the species are arranged in food chains which (9) combine to form a whole food cycle. This is closely bound up with the second principle, (10) the *Size of Food.* Although animals vary much in size, any one species of animal eats food only between certain limits of size, both lower and (11) upper, which (12) are illustrated by examples of a toad, a fly, and a bird. (13) This principle applied to primitive man, but no longer holds for civilized man, and (14) although there are certain exceptions to it in nature, it is a principle of great importance. (15) The third principle is that of *Niches.* By a niche is meant the animal's place in its community, its relations to food and enemies, and to some extent to other factors also. (16), (17), (18), (19), (20) A number of examples of niches can be given, many of which show that the same niche may be filled by entirely different animals in different parts of the world. (21) The fourth idea is that of the *Pyramid of Numbers* in a community, by which is meant the greater abundance of animals at the base of food chains, and the comparative scarcity of animals at the end of such chains. (22) Examples of this principle are given, but, as is the case with all work upon animal communities, good data are very scarce at present.

1. If you go out on to the Malvern Hills in July you will find some of the hot limestone pastures on the lower slopes covered with anthills made by a little yellow ant *(Acanthomyops flavus).* These are low hummocks about a foot in diameter, clothed with plants, some of which are different from those of the surrounding pasture. This ant, itself forming highly organized colonies, is the center of a closely knit community of other animals. You may find green woodpeckers digging great holes in the anthills, in order to secure the ants and their pupae. If you run up quickly to one of these places, from which a woodpecker has been disturbed, you may find that a robber ant *(Myrmica scabrinodis)* has seized the opportunity to carry off one of the pupae left behind by the yellow ants in their flight. The latter with

unending labor keep building up the hills with new soil, and on this soil there grows a special set of plants. Wild thyme *(Thymus serpyll)* is particularly common there, and its flowers attract the favorable notice of a red-tailed bumblebee *(Bombus lapidarius)* which visits them to gather nectar. Another animal visits these anthills for a different purpose: rabbits, in common with many other mammals, have the peculiar habit of depositing their dung in particular spots, often on some low hummock or tree stump. They also use anthills for this purpose, and thus provide humus which counteracts to some extent the eroding effects of the woodpeckers. It is interesting now to find that wild thyme is detested by rabbits as a food, which fact perhaps explains its prevalence on the anthills. There is a moth *(Pempelia subornatella)* whose larvae make silken tubes among the roots of wild thyme on such anthills; then there is a great army of hangers-on, guests, and parasites in the nests themselves; and so the story could be continued indefinitely. But even this slight sketch enables one to get some idea of the complexity of animal interrelations in a small area.

2. One might leave the ants and follow out the effects of the rabbits elsewhere. There are dor-beetles *(Geotrupes)* which dig holes sometimes as much as four feet deep, in which they store pellets of rabbit dung for their own private use. Rabbits themselves have far-reaching effects upon vegetation, and in many parts of England they are one of the most important factors controlling the nature and direction of ecological succession in plant communities, owing to the fact that they have a special scale of preferences as to food, and eat down some species more than others. Some of the remarkable results of "rabbit action" on vegetation may be read about in a very interesting book by Farrow. Since rabbits may influence plant communities in this way, it is obvious that they have indirectly a very important influence upon other animals also. Taking another line of investigation, we might follow out the fortunes and activities of the green woodpeckers, to find them preying on the big red and black ant *(Formica rufa)* which builds its nests in woods, and which in turn has a host of other animals linked up with it.

If we turned to the sea, or a fresh-water pond, or the

inside of a horse, we should find similar communities of animals, and in every case we should notice that food is the factor which plays the biggest part in their lives, and that it forms the connecting link between members of the communities.

3. In England we do not realize sufficiently vividly that man is surrounded by vast and intricate animal communities, and that his actions often produce on the animals effects which are usually quite unexpected in their nature—that in fact man is only one animal in a large community of other ones. This ignorance is largely to be attributed to town life. It is no exaggeration to say that our relations with the other members of the animal communities to which we belong have had a big influence on the course of history. For instance: the Black Death of the Middle Ages, which killed off more than half the people in Europe, was the disease which we call plague. Plague is carried by rats, which may form a permanent reservoir of the plague bacilli, from which the disease is originally transmitted to human beings by the bites of rat fleas. From this point it may either spread by more rat fleas or else under certain conditions by the breathing of infected air. Plague was still a serious menace to life in the seventeenth century, and finally flared up in the Great Plague of London in 1665, which swept away some hundred thousand people. Men at that time were still quite ignorant of the connection between rats and the spread of the disease, and we even find that orders were given for the destruction of cats and dogs because it was suspected that they were carriers of plague. And there seemed no reason why plague should not have continued indefinitely to threaten the lives of people in England; but after the end of the seventeenth century it practically disappeared from this country. This disappearance was partly due to the better conditions under which people were living, but there was also another reason. The dying down of the disease coincided with certain interesting events in the rat world. The common rat of Europe had been up to that time the Black or Ship Rat *(R. rattus)*, which is a very effective plague-carrier, owing to its habit of living in houses in rather close contact with man. Now, in 1727 great hordes of rats belonging to

another species, the Brown Rat *(R. norvegicus)*, were seen marching westwards into Russia, and swimming across the Volga. This invasion was the prelude to the complete occupation of Europe by brown rats. Furthermore, in most places they have driven out and destroyed the original black rats (which are now chiefly found on ships), and at the same time have adopted habits which do not bring them into such close contact with man as was the case with the black rat. The brown rat went to live chiefly in the sewers which were being installed in some of the European towns as a result of the onrush of civilization, so that plague cannot so easily be spread in Europe nowadays by the agency of rats. These important historical events among rats have probably contributed a great deal to the cessation of serious plague epidemics in man in Europe, although they are not the only factors which have caused a dying down of the disease. But it is probable that the small outbreak of plague in Suffolk in the year 1910 was prevented from spreading widely, owing to the absence of very close contact between man and rats. We have described this example of the rats at some length, since it shows how events of enormous import to man may take place in the animal world, without any one being aware of them.

4. The history of malaria in Great Britain is another example of the way in which we have unintentionally interfered with animals and produced most surprising results. Up to the end of the eighteenth century malaria was rife in the low-lying parts of Scotland and England, as also was liver-rot in sheep. No one in those days knew the causes or mechanisms of transmission of either of these two diseases; but at about that time very large parts of the country were drained in order to reclaim land for agricultural purposes, and this had the effect of practically wiping out malaria and greatly reducing liver-rot—quite unintentionally! We know now that malaria is caused by a protozoan which is spread to man by certain blood-sucking mosquitoes whose larvae live in stagnant water, and that the larva of the liver-fluke has to pass through one stage of its life history in a fresh-water snail (usually *Limnaea truncatula*). The existence of malaria depends on an abundance of mosquitoes, while that of liver-

rot is bound up with the distribution and numbers of the snail. With the draining of land both these animals disappeared or became much rarer.

5. On the whole, however, we have been settled in this country for such a long time that we seem to have struck a fairly level balance with the animals around us; and it is because the mechanism of animal society runs comparatively smoothly that it is hard to remember the number of important ways in which wild animals affect man, as, for instance, in the case of earthworms, which carry on such a heavy industry in the soil, or the whole delicately adjusted process of control of the numbers of herbivorous insects. It is interesting therefore to consider the sort of thing that happens when man invades a new country and attempts to exploit its resources, disturbing in the process the balance of nature. Some keen gardener, intent upon making Hawaii even more beautiful than before, introduced a plant called *Lantana camara,* which in its native home of Mexico causes no trouble to anybody. Meanwhile, someone else had also improved the amenities of the place by introducing turtle doves from China, which, unlike any of the native birds, fed eagerly upon the berries of *Lantana*. The combined effects of the vegetative powers of the plant and the spreading of seeds by the turtle doves were to make the *Lantana* multiply exceedingly and become a serious pest on the grazing country. Indian mynah birds were also introduced, and they too fed upon *Lantana* berries. After a few years the birds of both species had increased enormously in numbers. But there is another side to the story. Formerly the grasslands and young sugarcane plantations had been ravaged yearly by vast numbers of armyworm caterpillars, but the mynahs also fed upon these caterpillars and succeeded to a large extent in keeping them in check, so that the outbreaks became less severe. About this time certain insects were introduced in order to try to check the spread of *Lantana,* and several of these (in particular a species of Agromyzid fly) did actually destroy so much seed that the *Lantana* began to decrease. As a result of this, mynahs also began to decrease in numbers to such an extent that there began to occur again severe outbreaks of armyworm caterpillars. It was then found

that when the *Lantana* had been removed in many places, other introduced shrubs came in, some of which are even more difficult to eradicate than the original *Lantana*.

6. It is clear that animals are organized into a complex society, as complex and as fascinating to study as human society. At first sight we might despair of discovering any general principles regulating animal communities. But careful study of simple communities shows that there are several principles which enable us to analyze an animal community into its parts, and in the light of which much of the apparent complication disappears. These principles will be considered under four headings:

A Food chains and the food cycle.
B Size of food.
C Niches.
D The pryamid of numbers.

Food Chains and the Food Cycle

7. We shall see in a later chapter what a vast number of animals can be found in even a small district. It is natural to ask: "What are they all doing?" The answer to this is in many cases that they are not doing anything. All cold-blooded animals and a large number of warm-blooded ones spend an unexpectedly large proportion of their time doing nothing at all, or at any rate, nothing in particular. For instance, Percival says of the African rhinoceros: "After drinking they play . . . the rhino appears at his best at night and gambols in sheer lightness of heart. I have seen them romping like a lot of overgrown pigs in the neighborhood of the drinking place."

Animals are not always struggling for existence, but when they do begin, they spend the greater part of their lives eating. Feeding is such a universal and commonplace business that we are inclined to forget its importance. The primary driving force of all animals is the necessity of finding the right kind of food and enough of it. Food is the burning question in animal society, and the whole structure and activities of the community are dependent upon questions of food supply. We are not concerned here with the various devices employed by animals to enable them to obtain their food, or with the physiological processes which enable them to utilize in their tissues the energy derived from it. It is sufficient to bear in mind that animals have to depend ultimately upon plants for their supplies of energy, since plants alone are able to turn raw sunlight and chemicals into a form edible to animals. Consequently herbivores are the basic class in animal society. Another difference between animals and plants is that while plants are all competing for much the same class of food, animals have the most varied diets, and there is a great divergence in their food habits. The herbivores are usually preyed upon by carnivores, which get the energy of the sunlight at third-hand, and these again may be preyed upon by other carnivores, and so on, until we reach an animal which has no enemies, and which forms, as it were, a terminus on this food cycle. There are, in fact, chains of animals linked together by food, and all dependent in the long run upon plants. We refer to these as "food chains," and to all the food chains in a community as the "food cycle."

8. Starting from herbivorous animals of various sizes, there are as a rule a number of food chains radiating outwards, in which the carnivores become larger and larger, while the parasites are smaller than their hosts. For instance, in a pine wood there are various species of aphids or plant lice, which suck the juices of the tree, and which are preyed on by spiders. Small birds such as tits and warblers eat all these small animals, and are in turn destroyed by hawks. In an oak wood there are worms in the soil, feeding upon fallen leaves of plants, and themselves eaten by thrushes and blackbirds, which are in turn hunted and eaten by sparrow hawks. In the same wood there are mice, one of whose staple foods is acorns, and these form the chief food of the tawny owl. In the sea, diatoms form the basic plant food, and there are a number of crustacea (chiefly copepods) which turn these algae into food which can be eaten by larger animals. Copepods are living winnowing fans, and they form what may be called a "key industry" in the sea. The term "key industry" is a useful one, and is used to denote animals

which feed upon plants and which are so numerous as to have a very large number of animals dependent upon them. This point is considered again in the section of "Niches."

9. Extremely little work has been done so far on food cycles, and the number of examples which have been worked out in even the roughest way can be counted on the fingers of one hand.

Sometimes plants are not the immediate basis of the food cycle. This is the case with scavengers, and with such associations as the fauna of temporary fresh-water pools and of the abyssal parts of the sea, where the immediate basic food is mud and detritus; and the same is true of many parasitic faunas. In all these cases, which are peculiar, the food supply is of course ultimately derived from plants, but owing to the isolation of the animals it is convenient to treat them as a separate community.

Certain animals have succeeded in telescoping the particular food chain to which they belong. The whalebone whale manages to collect by means of its sievelike apparatus enough copepods and pteropods to supply its vast wants, and is not dependent on a series of intermediate species to produce food large enough for it to deal with effectively. This leads us on to a more detailed consideration of the problem of

Size of Food

10. Size has a remarkably great influence on the organization of animal communities. We have already seen how animals form food chains in which the species become progressively larger in size or, in the case of parasites, smaller in size. A little consideration will show that size is the main reason underlying the existence of these food chains, and that it explains many of the phenomena connected with the food cycle.

There are very definite limits, both upper and lower, to the size of food which a carnivorous animal can eat. It cannot catch and destroy animals *above* a certain size, because it is not strong or skillful enough. In the animal world, fighting weight counts for as much as it does among ourselves, and a small animal can no more tackle a large one successfully than a lightweight boxer can knock out a trained man four stone heavier than himself. This is obvious enough in a broad way: spiders do not catch elephants in their webs, nor do water scorpions prey on geese. Also the structure of an animal often puts limits to the size of food which it can get into its mouth. At the same time a carnivore cannot subsist on animals *below* a certain size, because it becomes impossible at a certain point to catch enough in a given time to supply its needs. If you have ever got lost on the moors and tried to make a square meal off bilberries, you will at once see the force of this reasoning. It depends, however, to a large extent on the numbers of the prey: foxes find it worthwhile to live entirely on mice in the years when the latter are very abundant, but prey on larger animals like rabbits at other times.

11. It is thus plain that the size of the prey of carnivorous animals is limited in the upward direction by its strength and ability to catch the prey, and in the downward direction by the feasibility of getting enough of the smaller food to satisfy its needs, the latter factor being also strongly influenced by the numbers as well as by the size of its food. The food of every carnivorous animal lies therefore between certain size limits, which depend partly on its own size and partly on other factors. There is an *optimum* size of food which is the one usually eaten, and the limits actually possible are not usually realized in practice. (It is as well to point out that herbivorous animals are not strictly limited by the size of their plant food, except in special cases such as seed-eating birds, honey-collecting insects, etc., owing to the fact that the plant cannot usually run away, or make much resistance to being eaten.) We have very little information as to the exact relative size of enemies and their prey, but future work will no doubt show that the relation is fairly regular throughout all animal communities.

12. Three examples will serve to illustrate the part played by size. There lives in the forests round Lake Victoria a kind of toad which is able to adjust its size to the needs of the moment. When attacked by a certain snake the toad swells itself out and becomes puffed up to such an ex-

tent that the snake is quite unable to cope with it, and the toad thus achieves its object, unlike the frog in Aesop's fable. Carpenter has pointed out another curious case of the importance of size in food. The tsetse fly *(Glossina palpalis)*, whose ecology was studied by him in the region of Lake Victoria, can suck the blood of many mammals and birds, in which the size of the blood corpuscles varies from 7 to 18μ, but is unable to suck that of the lungfish, since the corpuscles of the latter (41 μ in diameter) are too large to pass up the proboscis of the fly. A third case is that noticed by Vallentin in the Falkland Islands. He found that the black curlew *(Haematopus quoyi)* ate limpets *(Patella aenea)* on the rocks at low tide, but was able only to dislodge those of moderate size, not usually more than 45 millimeters across.

13. These are three rather curious cases of what is a universal phenomenon. Man is the only animal which can deal with almost any size of food, and even he has been able to do this only during the later part of his history. It appears that the very early ancestors of man must have eaten food of a very limited range of size—such things as shellfish, fruits, mushrooms, and small mammals. Later on, man developed the art of hunting and trapping large animals, and he was thus able to increase the size of his food in the *upward* direction, and this opened up possibilities of obtaining food in greater bulk and variety. After the hunting stage came the agricultural stage, and this consisted essentially in the further development of the use of large animals, now in a domesticated state, and in the invention of means of dealing with foods much *smaller* than had previously been possible, by obtaining great quantities of small seeds in a short time. All other animals except man have their food strictly confined within rather narrow limits of size. The whalebone whale can feed on tiny crustacea not a thousandth of its bulk, while the killer whale can destroy enormous cuttlefish; but it is only man who has the power of eating small, large, and medium-sized foods indiscriminately. This is one of the most important ways in which man has obtained control over his surroundings, and it is pretty clear that if other animals had the same power, there would not be anything like the same variety and specialization that there is among

them, since the elaborate and complex arrangements of the food cycles of animal communities would automatically disappear. For the very existence of food chains is due mainly to the fact that any one animal can live only on food of a certain size. Each stage in an ordinary food chain has the effect of making a smaller food into a larger one, and so making it available to a larger animal. But since there are upper and lower limits to the size of animals, a progressive food chain cannot contain more than a certain number of links, and usually has less than five.

14. There is another reason why food chains stop at a certain point; this is explained in the section on the Pyramid of Numbers. Leaving aside the question of parasites at present, it may be taken as a fairly general rule that the enemy is larger than the animal upon which it preys. (This idea is contained in the usual meaning of the word "carnivore.") But such is not invariably the case. Fierceness, skill, or some other special adaptation can make up for small size. The arctic skua pursues and terrorizes kittiwake gulls and compels them to disgorge their last meal. It does this mainly by naked bluff, since it is, as a matter of fact, rather less in weight than the gull, but is more determined and looks larger owing to a great mass of fluffy feathers. In fact, when we are dealing with the higher animals such as birds, mammals, and the social ants and bees, the psychology of the animals very often plays a large part in determining the relative sizes of enemies and their prey. Two types of behavior may be noticed. The strength of the prey and therefore its virtual size may be reduced; this is done by several devices, of which the commonest are poison and fear. Some snakes are able to paralyze and kill by both these methods, and so can cope with larger animals than would otherwise be possible. Stoats are able to paralyze rabbits with fear, and so reduce the speed and strength of the latter. It is owing to this that the stoat can be smaller than its prey. The fox, which does not possess this power of paralyzing animals with fear, is considerably larger than the rabbit. The second point is that animals are able to increase their own effective size by flock tactics. Killer whales in the Antarctic seas have been seen to unite in parties of three or four in order to break up

the thick ice upon which seals, their prey, are sleeping. Wolves are another example. Most wolves are about half the linear size of the deer which they hunt, but by uniting in packs they become as formidable as one very large animal. The Tibetan wolf, which eats small gazelles, etc., hunts singly or in twos and threes. On the other hand, herbivores often band together in flocks in order to increase their own powers of defense. This usually means increased strength, but other facts come in too. Ants have achieved what is perhaps the most successful solution of the size problem, since they form organized colonies whose size is entirely fluid according to circumstances. Schweitzer noted a column of driver ants in Angola march past for thirty-six hours. They are able by the mass action of their terrible battalions to destroy animals many times their own size (*e.g.*, whole litters of the hunting dog), and at the same time can carry the smallest of foods.

It must be remembered, therefore, that the idea of food chains of animals of progressively larger size is true only in a general way, and that there are a number of exceptions. Having considered the far-reaching effects of size on the organization of animal communities, we are now in a position to consider the subject of

Niches

15. It should be pretty clear by now that although the actual species of animals are different in different habitats, the ground plan of every animal community is much the same. In every community we should find herbivorous and carnivorous and scavenging animals. We can go further than this, however: in every kind of wood in England we should find some species of aphid, preyed upon by some species of ladybird. Many of the latter live exclusively on aphids. That is why they make such good controllers of aphid plagues in orchards. When they have eaten all the pest insects they just die of starvation, instead of turning their attention to some other species of animal, as so many carnivores do under similar circumstances. There are many animals which have equally well-defined food habits. A fox carries on the

very definite business of killing and eating rabbits and mice and some kinds of birds. The beetles of the genus *Stenus* pursue and catch springtails *(Collembola)* by means of their extensile tongues. Lions feed on large ungulates—in many places almost entirely zebras. Instances could be multiplied indefinitely. It is therefore convenient to have some term to describe the status of an animal in its community, to indicate what it is *doing* and not merely what it looks like, and the term used is "niche." Animals have all manner of external factors acting upon them—chemical, physical, and biotic—and the "niche" of an animal means its place in the biotic environment, *its relations to food and enemies.* The ecologist should cultivate the habit of looking at animals from this point of view as well as from the ordinary standpoint of appearance, names, affinities, and past history. When an ecologist says "there goes a badger" he should include in his thoughts some definite idea of the animal's place in the community to which it belongs, just as if he had said "there goes the vicar."

16. The niche of an animal can be defined to a large extent by its size and food habits. We have already referred to the various key-industry animals which exist, and we have used the term to denote herbivorous animals which are sufficiently numerous to support a series of carnivores. There is in every typical community a series of herbivores ranging from small ones (*e.g.*, aphids) to large ones (*e.g.*, deer). Within the herbivores of any one size there may be further differentiation according to food habits. Special niches are more easily distinguished among carnivores, and some instances have already been given.

The importance of studying niches is partly that it enables us to see how very different animal communities may resemble each other in the essentials of organization. For instance, there is the niche which is filled by birds of prey which eat small mammals such as shrews and mice. In an oak wood this niche is filled by tawny owls, while in the open grassland it is occupied by kestrels. The existence of this carnivore niche is dependent on the further fact that mice form a definite herbivore niche in many different associations, although the actual species of mice may be quite different. Or we might take as a niche all the carnivores which prey upon small mammals, and distinguish them from those which prey upon insects. When we do this it is immediately seen that the niches about which we have been speaking are only smaller subdivisions of the old conceptions of carnivore, herbivore, insectivore, etc., and that we are attempting only to give more accurate and detailed definitions of the food habits of animals.

17. There is often an extraordinarily close parallelism between niches in widely separated communities. In the arctic regions we find the arctic fox which, among other things, subsists upon the eggs of guillemots, while in winter it relies partly on the remains of seals killed by polar bears. Turning to tropical Africa, we find that the spotted hyena destroys large numbers of ostrich eggs, and also lives largely upon the remains of zebras killed by lions. The arctic fox and the hyena thus occupy the same two niches—the former seasonally, and the latter all the time. Another instance is the similarity between the sand martins, which one may see in early summer in a place like the Thames valley, hawking for insects over the river, and the bee-eaters in the upper part of the White Nile, which have precisely similar habits. Both have the same rather distinct food habits, and both, in addition, make their nests in the sides of sand cliffs forming the edge of the river valleys in which they live. (Abel Chapman says of the bee-eaters that "the whole cliff-face appeared aflame with the masses of these encarmined creatures.") These examples illustrate the tendency which exists for animals in widely separated parts of the world to drift into similar occupations, and it is seen also that it is convenient sometimes to include other factors than food alone when describing the niche of any animal. Of course, a great many animals do not have simple food habits and do not confine themselves religiously to one kind of food. But in even these animals there is usually some regular rhythm in their food habits, or some regularity in their diverse foods. As can be said of every other problem connected with animal communities, very little deliberate work has been done on the subject, although much information can be found in a scattered form, and only awaits careful coordination in order

to yield a rich crop of ideas. The various books and journals of ornithology and entomology are like a row of beehives containing an immense amount of valuable honey, which has been stored up in separate cells by the bees that made it. The advantage, and at the same time the difficulty, of ecological work is that it attempts to provide conceptions which can link up into some complete scheme the colossal store of facts about natural history which has accumulated up to date in this rather haphazard manner. This applies with particular force to facts about the food habits of animals. Until more organized information about the subject is available, it is possible only to give a few instances of some of the more clear-cut niches which happen to have been worked out.

18. One of the biggest niches is that occupied by small sapsuckers, of which one of the biggest groups is that of the plant lice or aphids. The animals preying upon aphids form a rather distinct niche also. Of these the most important are the coccinellid beetles known as ladybirds, together with the larvae of syrphid flies (cf. Figure V-6) and lacewings. The

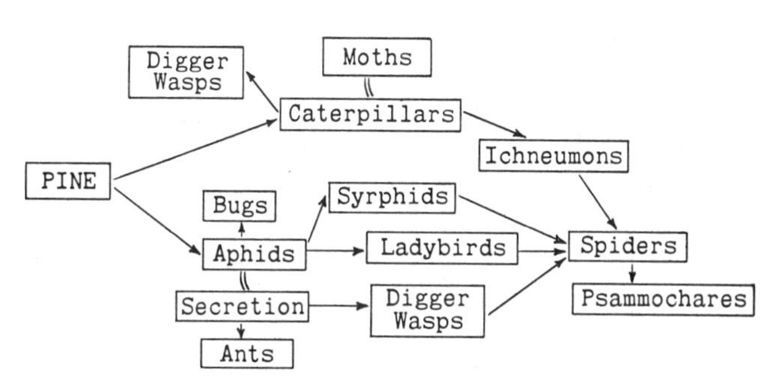

V-6. *Food cycle on young pine trees on Oxshott Common. (From Richards.)*

niche in the sea and in fresh water which is analogous to that of aphids on land in filled by copepods, which are mainly diatomeaters. This niche occurs all over the world, and has a number of well-defined carnivore niches associated with it.

If we take a group of animals like the herbivorous grass-eating mammals, we find that they can be divided into smaller niches according to the size of the animals. There is the mouse niche, filled by various species in different parts of the world; the rabbit niche, of larger size, filled by rabbits and hares in the palaearctic region and in North America, by the agouti and viscacha in South America, by wallabies in Australia, and by animals like the hyrax, the springbuck, and the mouse deer in Africa. In the same way it can be shown that there is a special niche of carnivorous snakes which prey upon other snakes—a niche which is filled by different species in different countries. In South America there is the mussarama, a large snake four or five feet in length, which is not itself poisonous, but preys exclusively upon other snakes, many of which are poisonous, being itself immune to the venoms of lachesis and rattlesnake, but not to colubrine poisons. In the United States the niche is filled by the king snake which has similar habits, while in India there is a snake called the hamadryad which preys upon other (in this case nonpoisonous) snakes.

19. Another widespread niche among animals is that occupied by species which pick ticks off other animals. For instance, the African tick bird feeds entirely upon the ticks which live upon the skin of ungulates, and is so closely dependent upon its mammalian "host" that it makes its nest of the latter's hair (*e.g.* of the hartebeest). In England, starlings can often be seen performing the same office for sheep and deer. A similar niche is occupied on the Galapagos Islands by a species of scarlet land crab, which has been observed picking ticks off the skin of the great aquatic lizards *(Amblyrhynchus).* Another niche, rather analogous to the last one, is that occupied by various species of birds, which follow herds of large mammals in order to catch the insects which are disturbed by the feet of the animals. Chapman saw elephants in the Sudan being followed by kites and grey herons; Percival says that the buff-backed egret follows elephants and buffalo in Kenya for the same purpose; in Paraguay there are the Aru blackbirds which feed upon insects disturbed by the feet of cattle; while in England wagtails attend cattle and sheep in the same way.

20. There is a definite niche which is usually filled by earthworms in the soil, the species of worm differing in different parts of the world. But on coral islands their place may be largely taken by land crabs. Wood-Jones states that on Cocos-Keeling Island, coconut husks are one of the most important sources of humus in the soil, and in the rotting husks land crabs (chiefly of the genus *Cardiosoma)* make burrows and do the same work that earthworms do in our own country. (There are as a matter of fact earthworms as well on these islands.) On the coral reefs which cover such a large part of the coast in tropical regions, there is a definite niche filled by animals which browse upon the corals, just as herbivorous mammals browse upon vegetation on land. There are enormous numbers of holothurians or sea cucumbers which feed entirely in this way. Darwin gives a very good description of this niche. Speaking also of Cocos-Keeling Island, he says:

> The number of species of Holothuria, and of the individuals which swarm on every part of these coral reefs, is extraordinarily great; and many shiploads are annually freighted as is well known, for China with the trepang, which is a species of this genus. The amount of coral yearly consumed, and ground down into the finest mud, by these several creatures, and probably by many other kinds, must be immense. These facts are, however, of more importance in another point of view, as showing us that there are living checks to the growth of coral-reefs, and that the almost universal law of "consume and be consumed," holds good even with the polypifers forming those massive bulwarks, which are able to withstand the force of the open ocean.

This passage, besides showing that the coral-eating niche has a geological significance, illustrates the wide grasp of ecological principles possessed by Darwin, a fact which continually strikes the reader of his works. We have now said enough to show what is meant by an ecological niche, and how the study of these niches helps us to see the fundamental similarity between many animal communities which may appear very different superficially. The niche of an animal may to some extent be defined by its numbers. This leads us on to the last subject of this chapter.

The Pyramid of Numbers

21. "One hill cannot shelter two tigers." In other and less interesting words, many carnivorous animals, especially at or near the end of a food chain, have some system of territories, whereby it is arranged that each individual, or pair, or family, has an area of country sufficiently large to supply its food requirements. Hawks divide up the country in this way, and Eliot Howard's work has shown that similar territory systems play a very important part in the lives of warblers. We can approach the matter also from this point of view: the smaller an animal the commoner it is on the whole. This is familiar enough as a general fact. If you are studying the fauna of an oak wood in summer, you find vast numbers of small herbivorous insects, like aphids, a large number of spiders and carnivorous ground beetles, a fair number of small warblers, and only one or two hawks. Similarly in a small pond, the numbers of protozoa may run into millions, those of *Daphnia* and *Cyclops* into hundreds of thousands, while there will be far fewer beetle larvae, and only a very few small fish. To put the matter more definitely, the animals at the base of a food chain are relatively abundant, while those at the end are relatively few in numbers, and there is a progressive decrease in between the two extremes. The reason for this fact is simple enough. The small herbivorous animals which form the key industries in the community are able to increase at a very high rate chiefly by virtue of their small size), and are therefore able to provide a large margin of numbers over and above that which would be necessary to maintain their population in the absence of enemies. This margin supports a set of carnivores, which are larger in size and fewer in numbers. These carnivores in turn can provide only a still smaller margin, owing to their large size which makes them increase more slowly, and to their smaller numbers. Finally, a point is reached at which we find a carnivore (*e.g.*, the lynx or the peregrine falcon) whose numbers are so small

that it cannot support any further stage in the food chain. There is obviously a lower limit in the density of numbers of its food at which it ceases to be worthwhile for a carnivore to eat that food, owing to the labor and time that is involved in the process. It is because of these number relations that carnivores tend to be much more wide-ranging and less strictly confined to one habitat than herbivores.

22. This arrangement of numbers in the community, the relative decrease in numbers at each stage in a food chain, is characteristically found in animal communities all over the world, and to it we have applied the term "pyramid of numbers." It results, as we have seen, from the two facts (*a*) that smaller animals are preyed upon usually by larger animals, and (*b*) that small animals can increase faster than large ones, and so are able to support the latter.

The general existence of this pyramid in numbers hardly requires proving, since it is a matter of common observation in the field. Actual figures for the relative numbers of different stages in a food chain are very hard to obtain in the present state of our knowledge. But three examples will help to crystallize the idea of this "pyramid." Birge and Juday have calculated that the material which can be used as food by the plankton rotifers and crustacea of Lake Mendota in North America weighs twelve to eighteen times as much as they do. (The fish which eat the crustacea would weigh still less.) Again, Mawson estimated that one pair of skuas (*Megalestris*) on Haswell I. in the Antarctic regions, required about fifty to one hundred Adelie penguins to keep them supplied with food (in the form of eggs and young of the penguins); while Percival states that one lion will kill some fifty zebras per year, which gives us some idea of the large numbers of such a slow-breeding animal as the zebra which are required to produce this extra margin of numbers.

The relationship of predator and prey is one of the more obvious strands which bind the animal community. But there are others less immediately apparent—among them parasitism, symbiosis, and commensalism—which offer a wide field for nature study. The relationship of shark and remora or of human being and tapeworm are relatively familiar. But there are others of such diverse and fantastic forms that they exert a fascination on both ecologist and layman. Some of them are described in this selection from The Sociology of Nature, *by Leslie Reid, a professional forester, teacher, and author of numerous articles on geological and biological subjects.*

STRANDS OF DEPENDENCE

LESLIE REID

MOST OF US feel that though the relationship between predator and prey is savage and relentless, it is something we can accept, if for no other reason than that we practice it ourselves. That between parasite and host, on the other hand, we regard as repulsive, far less easily acceptable to fastidious minds than any other of nature's ways. This is an understandable prejudice, but like all prejudices basically illogical. We are deeply moved by the beauty of living creatures. This beauty takes many forms, the most noticeable being of color and shape. Another and almost equally important is that of adaptation to circumstances of their lives. The bumblebee exploring the flowers of a foxglove spire, a swift cleaving the air with a scimitar of black wings, the Indian leaf butterfly, with the underside of its wings counterfeiting a leaf not only in shape, but in markings as well; all these and hundreds of others are miracles of adaptation and are beautiful for precisely that reason. But what could be more of a miracle of adaptation than a tapeworm which has degenerated to little more than a set of hooks and a series of detachable segments, each containing eggs by the thousand. Sightless, immobile, living in total darkness and bathed perpetually in its food which it absorbs through the skin, it has no need for any of the attributes it has forsworn. Alternatively consider the swift mentioned above. This bird, and particularly the nestling, is infested with anything up to twelve louse flies, flightless, bloodsucking, and of repulsive aspect, each one as big in proportion to its host as a moderate-size crab would be to a human body. The louse fly produces a single larva which pupates at once and survives the winter in or near the nest. From this pupa an adult emerges, timing its emergence with the hatching of the swift's eggs in June. Adaptation could scarcely go further.

There is another point of view which should make us hesitate to accuse parasites of degenerate, still less of reprehensible, behavior. To do so is to regard the natural world from a purely human standpoint. Who are we to pass judgment? It is of course true that there is a big difference between an animal that devours another and one that lives on or in a permanent host; and this difference, as Charles Elton has vividly pointed out, is that between living on capital, which is the way of the carnivore, and living on income, which is the way of the parasite. Foxes eat rabbits, and in so doing destroy them so that they are no longer capable of providing food; but the tapeworm inhabiting a rabbit goes one better and induces the rabbit to supply it with sustenance for as long as the rabbit continues to live. The resemblance is really more important than the difference, and this is made clearer still by the fact that, as so often in nature, there is no sharply dividing line between the one way of life and the other. The tapeworm is a confirmed and dedicated parasite, its host the rabbit a free-living creature. But in between these extremes there are many connecting links. The louse, like the tapeworm, is a parasite but less completely so, since it will remove itself from one host to another. The flea goes further and spends much of its time wandering abroad. The warble fly is a parasite for its larval stage only, while as for the bloodsucking flies, it is not easy to say whether they are carnivores or parasites.

Parasitism, whether we like it or not, is a widespread and highly successful way of life. It is a way of life moreover subject to rules and restrictions like any other. Complete parasites, that is to say those that pass the whole or the major part of their lives on or in their host, have a really remarkable achievement to their credit, for while they are predators they have succeeded in converting their prey into their environment. With them habitat and food are one and the same. It is for this reason, if for no other, that they are very much the concern of the ecologist. As for the general success of parasites, we have only to consider that, for instance a bird, almost any kind of bird, is quite likely to harbor some twenty different kinds of parasite in all parts of its body, internally and externally—protozoa and bacteria in the bloodstream; worms of various kinds, flukes, tapeworms, roundworms, leeches, in the digestive system, in the bronchial tubes, in the air sacs and lungs; bugs, fleas, feather lice, mites and ticks, infesting the plumage and the skin. The number of parasites of one kind or another is sometimes past belief. Over ten thousand nematode worms, for instance, have been taken from a single grouse.

Some parasites, such as many of the feather lice of birds, are host-specific, which means that they infest one kind of bird and one only. So frequently does this happen that they have been made use of to trace the descent of certain closely related species of bird. As a way of living, parasitism varies in the closeness of physical intimacy between parasitic associations of two free-living animals at one extreme, and actual physiological union at the other. Of the first group there are several examples. One is the pirate or klepto-parasite, such as the skua that harries gulls until they disgorge the fish they have swallowed. Then there is the brood-parasite like our familiar cuckoo whose evasion of parental responsibilities is well known, or the ichneumon fly that lays her eggs in the living bodies of certain caterpillars. There are parasitoids like the hunting wasps to whom the host is a fly or a spider which serves as food for the larvae of the wasp. Of the second group, involving the most intimate forms of parasitism, perhaps the strangest example is found in some of the deep-sea angler fish, among whom parasitism is a matter of sex. The depths of the sea constitute an environment so enormous as to be sparsely settled. This means that, especially with rather lethargic lurkers like these angler fish, it is no easy matter for male and female to meet. Once met it is advisable never to part. The male, the more active of the two, is little more than a quarter as big as the female. He probably tracks her down by some chemical means, and having found her attaches himself in so intimate a manner that the bloodstreams of the two actually unite.

What are the rules of the parasite game? There is one golden one, to work out a compromise between drawing all possible sustenance from the host on the one hand, and on the other grievously impairing or even killing it in the process.

The flow of income must be at once adequate and perennial while life endures. Are the odds entirely against the host? It is true that as a rule it seems to come to no permanent harm, and can hit back to some slight extent by preening, dust-bathing, scratching or biting, but such measures are of use only against ectoparasites (those parasites that live on the outside of the host). There is little that the host can do against the sheltered endo-parasite, except tolerate it.

PARASITISM AND ECOLOGY. Since, as I have said, to the parasite its host is also its habitat, there is no aspect of the subject that can fail to be of interest to the ecologist. There are two above all that are his concern. One is the response of the parasite to its environment, in other words the adaptations consequent upon so specialized a mode of life. To study these is indeed to enter a bizarre world.

In the first place it must be understood that parasitism offers a secure and comfortable existence once it has been established. It would hardly be possible to overemphasize the weightiness of the qualification in that sentence for, as though to exact payment for easy living, the road leading to the desired goal is frequently obstructed with unbelievable hazards. Outstanding among these is the mathematical improbability of the larvae ever finding the right host. Consider for instance the roundworm that infests grouse. The eggs are scattered over the heather-covered hills, and when they hatch the larvae crawl to the tips of the heather-shoots, which are the food of the grouse. If one of these harboring a roundworm larva is eaten by one of the birds, all is well from the point of view of the parasite. If this does not happen the larva perishes. The odds against any of them must be of the order of millions to one.

The same sort of astronomical mortality must apply to ticks in tropical grasslands and leeches in the equatorial forest. As a result parasites generally possess a correspondingly astronomical fertility. They must lay eggs on an enormous scale and their reproductive organs must develop in order to do so. But that is not enough, and we find among parasites, as nowhere else to the same extent, a number of highly specialized reproductive devices. Consider first the problem of those in which the sexes are in different individuals, which means that two must enter the body of the host and these two of opposite sexes. The odds are heavy against them, and there are further heavy odds against the two meeting. To do so is a triumph and the most must be made of it. As with the angler fish already referred to, they must never be allowed to separate, and so in one kind of fluke, for instance, the male embeds the female in a groove in its side; and this condition, approximating to that of a parasite on a parasite, prevails until the eggs are fertilized. Parthenogenesis, or virgin birth, is a practice fully recognized, to such effect indeed that among some of the nematodes or roundworms no males have ever been found. On occasions nonsexual forms of reproduction are added to sexual ones. Thus the fertilized egg of a trematode or fluke may split up within the body of its intermediate host into several million larvae. Some tapeworms multiply simply by budding in the larval stage, or by giving off chains of individuals when adult.

Quite apart from these reproductive adaptations are certain characteristic structural ones. This is largely a matter of atrophy, of dispensing with parts of the body for which a parasite has no use. Limbs were among the first to go. All that many of them need are hooks or spines for hanging on. Consequently feather lice, bugs, and fleas have either lost their wings, or retain them only as useless vestiges, while their mouth parts are equipped with spines, often recurved. Leeches have suckers and exude a saliva which prevents coagulation of the blood of their victim. Many parasites have no need for either mouth parts or a digestive tract, since nutriment is absorbed through the surface of the body.

Eyes and ears are equally unnecessary for liver flukes or feather mites which have instead a specialized sense or tropism to guide them unerringly to the required part of the host's body.

LIFE CYCLES. It must be remembered that parasites, having, as I have said, converted their food supply into their environment, are in fact closely bound up with two environments, that of their host's body and that much wider one to which the host belongs. Feather lice and mites pass successive generations on the same victim, and are not called upon to face the hazards of the outside world. Most parasites are less fortunate, having larval stages, when for a time they must fend for themselves and become free-living creatures in a hostile world. As soon as they are eaten by the appropriate animal their parasitic adult stage begins. Here then is a relationship with the habitat of the outer scene, and perhaps it is not surprising that a great many parasites have learned to take advantage of the food-chain system, in truth of the whole food cycle governing the animal community of which they are members. This means distributing the stages of their life cycle among two or more hosts, one stage devoted to each. The advantage gained is an enhanced security, a reduction in some measure at least of the extreme chanciness of their existence. In the simpler instances it solves the problem of what happens to the parasite when its host is devoured by a predator, since an essential part of the scheme is that very act of devouring.

To take one of these simpler examples: foxes eat rabbits, and there is a tapeworm infesting both, passing the larval stage in the rabbit and the adult stage in the fox. The cycle is completed when the tapeworm in the fox lays enormous numbers of eggs which pass out with the faeces. These eggs infect the grass eaten by rabbits. The more subtle refinements of this process must not be lost sight of. The larvae and the adult tapeworms do not content themselves merely with entering the bodies of rabbit and fox respectively. The larvae embed themselves in the muscles of the rabbit which the fox can hardly fail to devour, while the adults settle down to a life of ease in the intestine of the fox, from whence the eggs will duly be extruded.

One more point must be made, again showing how parasitism links up with ecology. Parasites, apart from pirates and robbers, are necessarily much smaller than their hosts. Many of them are themselves parasitized. Ticks and fleas, for instance, have protozoan parasites of their own. Once more they must be smaller than their hosts. Here then is a food chain among parasites and hosts, but the pyramid of numbers, instead of being made up of smaller numbers of larger creatures, is here inverted and consists of a larger number of smaller creatures. When the complete food cycle is worked out showing one or more free-living creatures, together with their parasites and including the parasites on the parasites, some idea of the complexity of these strands of dependence becomes clear.

To make a somewhat paradoxical statement, if it can be shown that an assemblage of living creatures, in spite of being in keen and often bitter competition, have so organized their way of living as to restrict competition as far as possible, with the result that the resources of their environment are in effect shared, then that way of living deserves to be called cooperative. The food cycle, even though it means ruthless predation, is a method of portioning out available food. So, even more, is the system of ecological niches, according to which each species, and each individual within the species, can be reasonably certain of a particular kind of food without having to fight to obtain it. The territory system can be similarly described, resulting as it does at least in some measure of spacing out and avoidance of overcrowding. Where competitive dependence is concerned therefore we find that competition is cut down to a minimum, and this can be achieved only by a truly cooperative general system.

The Urge to Associate

This is a great deal, but is very far from being all. Cooperation among animals goes much farther, as far in fact as to justify the conclusion that there exists throughout the world of animals an innate and widespread tendency to cooperate, to organize their existence on a basis of mutual aid. This is nothing less than a fundamental ecological principle, and a great weight of evidence can be adduced in support of it.

Some kinds of association are highly specialized and intimate, involving animals of widely divergent relationship, a few, and these perhaps the most interesting of all, extending beyond the boundaries of the animal kingdom to include plants, intimate partnerships between an animal and a plant, or between two plants. Two kinds of partnership of this kind are distinguished, though it is not always easy to maintain the distinction. They are symbiosis and commensalism.

SYMBIOSIS. This is much the more intimate of the two, but the distinction between symbiosis and commensalism is usually drawn, not according to intimacy but according to whether benefit is mutual or one sided. In symbiosis it is mutual. Where two plants are concerned the most notable example is the lichens, whose texture is a close weaving of the threads of a fungus with the green cells of an alga. The fungus, like all fungi, is deficient in chlorophyll but derives carbohydrates from the alga, which does possess chlorophyll and so makes them from carbon dioxide by photosynthesis. In return for this the fungus provides the alga with water containing nutrient salts. Lichens are of the utmost interest for the animal ecologist because of the part they play as the earliest colonists of a stretch of bare rock, helping to convert it into a habitat for the higher plants, and so for animals. Then there are the nitrogen-fixing bacteria, partners in a symbiotic association with leguminous plants such as peas, beans, and clovers. The bacteria, in nodules on the roots, build up nitrogen compounds from the air and these are used by the plants. As their share of the bargain, the bacteria draw carbohydrates which the plants have manufactured by photosynthesis.

As for symbiotic partnerships between animals and plants, they are to be found in more than one of the provinces into which the animal kingdom is divided. Again the plant concerned is a unicellular alga, and the animals are green for the same reason that plants are green, that they contain these cells provided with cholorophyll. A minute flatworm known as *Convoluta* appears at low tide over stretches of sand on the northern coast of Brittany in such profusion as to color the sand like grass. They rise to the surface in their millions to enlist the cooperation of sunlight in their work of photosynthesis. When the tide begins to turn they submerge, and soon not a trace of verdure is to be seen. Mutual benefits received in this case are of more than one kind. The animal obtains oxygen from the photosynthesis of the plant, and at the same time receives assistance in excretion which gives rise to phosphates. These substances, which are waste products from the animal's point of view, are used by the plant to form protein. Another benefit derived by the plant is a supply of carbon dioxide, which is a product of the animal's respiration, and to this can be added the protection it enjoys from its close physiological association with the animal.

A very strange example of symbiosis is found among many of the termites, which subsist on a diet of wood. This seems to us an unsatisfying food, and so it is even to the termites, which are incapable of digesting it without the cooperation of flagellate protozoa living what appears to be a parasitic existence in their digestive tracts. The protozoa render the wood available both to themselves and to the termites, who if deprived of their assistance, as can be done by applying heat, die within twenty days.

A not uncommon association between animals is when one lives and grows upon another, and the smaller partner is then known as an epizoite, just as a plant growing upon another plant is called an epiphyte. This may lead to parasitism, but an epizoite lives on its own resources and not at the expense of its host. It is a common form of association in the sea and in fresh water, and some instances may well be symbiotic. In the Indian Ocean there lives a small rock perch, *Minous inermis,* which never seems to be free from a heavy infestation of hydroids. It looks as though the hydroids help to conceal the perch and in so doing achieve a mobility denied to most of their kind. A delightful and very remarkable refinement of the symbiotic way of life has been perfected by a crab of the genus *Melia,* again from the

Indian Ocean. This apparently sagacious creature has taken to plucking certain small sea anemones from the sand and swimming about with one of them clasped in each of its pincers. When the anemones catch food, after the manner of their kind, from the surrounding water, *Melia* extends one of its first pair of walking legs and removes what it considers to be its share. If this is indeed an instance of symbiosis, we may suppose that with the anemones the advantage is that gained by transport.

Passing to dry land, we find a number of associations among larger animals in which benefit appears to be mutual. Most of us have seen starlings perched on the backs of cattle and sheep. If, as seems highly probable, their purpose is to remove keds and lice, then this is a simple symbiotic relationship, the starlings gaining food and in doing so helping to relieve the sheep of the parasites that afflict them. There is no doubt about similar services rendered by the ox-pecker birds of the African savanna to the large grazing animals on whose backs they spend most of their time, running about all over them like mice, subsisting it seems entirely on ticks, biting flies and lice. It is said that they depend also for their nesting material on the coarse hair of their hosts, who clearly welcome their attentions, though it appears that there are occasions when the birds peck at wounds in the buffaloes' hide, and in so doing become parasites rather than symbionts.

No fewer than twenty-six species of fish are known to devote most, though not all of their time, to the systematic cleaning of other fish which welcome and eagerly seek out their services. Six species of shrimp do the same and at least one crab. The services of cleaners are so highly valued that predatory fish leave them alone, and in consequence there have arisen noncleaning impostors who have found it worthwhile to imitate them and so share the security they enjoy.

COMMENSALISM. The word means feeding together, and it is commonly stated that one of the partners benefits and not the other. The following examples illustrate this point, if on occasions rather doubtfully, while some multiple associations combine examples of symbiosis, commensalism and parasitism. As such they may be inconvenient for the systematist, but are of particular interest for the fact that they show this combination.

Consider for instance the well-known case of the hermit crab which has soft and very vulnerable hinder parts. These it protects by inhabiting a cast-off whelk shell. Frequently the shell, thus serving a secondary phase of usefulness, is found to be densely encrusted with polyps of the hydroid *Hydractina*. This could be construed as symbiosis, since the hydroids gain transport, perhaps protection and an enhanced food supply, while the crab is camouflaged by the growth on its shell.

As another instance of the difficulty of making the convenient and rigid distinctions beloved of the human mind, the very remarkable life history of the large blue butterfly, *Maculinea arion*, can be cited. This is one of the most astonishing of all animal associations, among other reasons because the larva of the butterfly is free-living for part of its existence, commensal and in a distinct sense parasitic for the rest. The large blue lays its eggs on wild thyme in a downland habitat, and the larva feeds on the thyme, living on its own up to the time of its third moult. After that a change comes over it. It begins to wander about as though looking for something. So it is. It is looking for an ant, having reached a crisis in its life of such urgency that if it fails to find an ant it will perish. At the same time it now possesses an inducement for an ant to play its part in the strange sequence of events that follow, for it has by this time developed a honey gland secreting a sweet fluid of the sort for which ants have a passion. Ants are by no means uncommon on the chalk downs, and the likelihood of the encounter taking place is considerable, especially as there is evidence for supposing that the butterfly usually lays her eggs at no great distance from an ant's nest.

As soon as the meeting takes place, the ant is sure to take a lively interest in the caterpillar, and the interest is welcomed. The caterpillar makes this clear at first by allowing itself to be milked, later by hunching itself up into a curious attitude, as though asking to be picked up. The ant responds as it was meant to, and presently carries the caterpillar off to its nest. It is in the darkness of this alien nest that the rest of the metamorphosis is worked out. Up to the time of its winter hibernation and afterwards, until it pupates, the larva feeds solely on the grubs of its hosts, the ants, who display not the least resentment, apparently considering themselves amply paid by the sweet liquid of the honey gland. This comes very close to symbiosis, though the caterpillar seems to have the best of the bargain. In the following summer this strange messmate, having emerged from the pupa, emerges also from the ant's nest to the open air, where its wings expand and it becomes a free-living butterfly.

Remaining among the ants we come upon another association of this commensal sort, and one that involves something much more than mere commensalism. The larva of the large blue butterfly is only one of the creatures that secretes a sweet fluid irresistible to ants. Notable among these are the aphids or greenfly, and it has been known for many years that these have come to mean so much to certain ants, among them the small black or garden ant very common in this country, that they have become pastoralists, solicitously tending herds of aphids that are regularly milked and put out to pasture, like so many cows, on suitable plants not far from the nest. Some kinds of aphids are kept in special compartments of the nest underground, where they subsist on the sap drawn from the roots of grasses. Some writers have made out that the benefit derived by aphids from this protection has saved them from extermination. Some degree of protection is not to be denied, and aphids are defenceless and extremely sluggish creatures, preyed upon to a devastating extent, but we are probably on surer ground if we attribute their survival to their prodigious powers of reproduction.

There are examples of commensalism among vertebrates. A remarkable one is that of a certain South American parakeet, which builds its nest and raises its brood in the carton nests of a species of termite and will do so in no other situation. When the birds arrive and begin to hollow out the termitary, the termites rush out to repair the damage, but quite soon change their tactics and decide to leave the birds alone. It would be wholly within their power to attack the nestlings subsequently, but this temptation too is overcome. In this instance it is clear enough what the parakeets stand to gain, since egg thieves are likely to respect termites. What the termites gain, if anything at all, it is impossible to say.

Again among vertebrates, is the well-known association between blue sharks and pilot fish. Some of the stories concerning them are no doubt legendary, but the association exists. A more intimate one is that between sharks and remoras, which are equipped with suckers for their very firm attachment to their hosts. This suggests parasitism, but there seems to be no evidence that the remora derives nourishment directly from the shark. We may suppose that it gains transport and perhaps scraps from time to time from the host's table. The shark on the other hand seems to gain nothing.